Total Productive Facilities Management

Richard W. Sievert, Jr.

Total

Productive

Facilities

Management

A Comprehensive Program To:
- Achieve Business Goals by Optimizing Facility Resources
- Implement Best Practices Through Benchmarking, Evaluation & Project Management
- Increase Your Value to the Organization

Richard W. Sievert, Jr.

RSMeans

Copyright 1998
R.S. Means Company, Inc.
Construction Publishers & Consultants
63 Smiths Lane
Kingston, MA 02364-0800
(781) 422-5000
www.rsmeans.com

Managing Editor. Mary Greene Editors Mary Greene and Phillip R. Waier, P E
Production Manager· Michael Kokernak. Production Coordinator Marion Schofield
Composition. Paula Reale-Camelio. Book and cover design: Norman R. Forgit

10 9 8 7 6 5 4

Library of Congress Catalog Number 98-162139

ISBN 0-87629-500-6

This book is dedicated to those who have preceded
me in leading our family business.

My great-grandfather, the late William J. Sievert—
    Company founder, 1917

My grandfather, the late Bernard E. Sievert—President, 1945 to 1966

My father, Richard Sievert—CEO, 1966 to present

Table of Contents

Foreword by Charles M. Boyles, C.P.E.   xi

Acknowledgments   xiii

Part I: Evaluating Facilities Performance and Setting Goals   1

  Chapter 1: Where Do You Stand Today?   3

    Total Productive Facilities Management (TPFM)   4

    Definition of Facility Management   5

    The Changing Facility Management Profession   6

    Staying Competitive While Controlling Costs   9

    The Facility Manager's Challenge   10

    Evaluating Your Management Approach   11

    Planning and Organizing a Program to Achieve
Continuous Improvement   11

    Know Your Customers   13

    Know Your Environment   15

    Summary   20

  Chapter 2: The Need for Facilities Performance Evaluations   23

    Standards   23

    Types of Facility Evaluations   25

    Summary   32

  Chapter 3: Increasing Productivity Through Benchmarking   39

    Statistical Process Control   41

    Occupant Survey Method   44

    Ratio Analysis Method   44

    Value Engineering Studies   44

    The Benchmarking Process   46

Waste Minimization   59

Environmental Health and Safety Benchmarks   76

Summary   82

Chapter 4: Value Engineering: Doing More With Less   85

Definition of Value Engineering   86

Value Engineering Applications in Facility Management   86

The Value Engineering Job Plan   89

Value Engineering Change Proposal   99

Other Techniques to Lower Cost and Improve Value   102

Value Engineering: Past, Present, and Future   102

Part II: Project-by-Project Improvement   105

Chapter 5: The Project Management Process   107

Definition of Project Management   107

Characteristics of Projects   108

Creating a Project-Driven Facility Management Organization   108

Outsourcing and Out-Tasking   108

Key Aspects of Project Success   112

Project Organization Structures   112

A Hybrid Organization Structure   120

The Project Life Cycle   124

Role of the Project Manager   126

How to Plan and Scope Projects   128

Work Breakdown Structure   130

Communicating the Scope of the Work   131

Summary   135

Chapter 6: Cost Management   141

Estimating Project Costs   143

Project Budget Checklist   146

Cost Information Sources   147

Monitoring and Reporting Costs   148

Value Engineering   148

Cost Trade-Off Studies   148

Building Economics   149

Life Cycle Cost Analysis   151

Cash Flow Elements   151

Time Value of Money   152

Life Cycle Cost Analysis Output/Results   153

Summary   157

Chapter 7: Schedule Planning and Control   159

    Project Scheduling Methods   159

    Milestone Charts   160

    Bar Charts   161

    Network Scheduling   163

    Critical Path Method (CPM)   165

    Balancing Project Time and Cost Relationships (Resource Leveling)   175

    Summary   175

Chapter 8: Contracting and Procurement Methods   181

    General Contractor   181

    Design/Build   184

    Construction Management   186

    Contract Documents   191

    Invitation to Bid   194

    Potential Owner Liabilities   199

    Job Site Safety Programs   200

    Insurance Protection   201

    Methods of Compensation   202

    Creative Ways to Finance and Contract New Projects   204

    Summary   205

Part 3: Improved Productivity Through Communications and Teamwork   207

Chapter 9: Good Communication: Vital to Teamwork   209

    Project Communications   211

    Written Communications   212

    Meetings, Meetings, Meetings   213

    Project Record Filing System   213

    Handling Change Orders   218

    On-Site Communication   218

    A Word About Advanced Communications Technology   219

    Summary   220

Chapter 10: Team Problem-Solving and Decision-Making Techniques   221

    Assembling of Contractors and Consultants for the Project Team   221

    Techniques for Team Decision-Making and Problem-Solving   222

    Post-Mortem Team Review   226

    Characteristics of Team Leaders   227

Epilogue: Where Do You Go from Here?   231

Appendix   237

Bibliography/References   249

Index   257

Foreword

Successful facility management is not a function of unexplained, unpredictable, and random events. Nor is the failure of any facility to perform successfully caused by random chaos. According to Mr. Sievert, successful facility management depends on an in-depth understanding of the purpose, function, customer or user of the facility, and the life span of the facility.

Contemporary economics demand accurate and valid assessments of facilities. The methodology presented in the following pages offers readers an atlas for this key function in successful facility management. Drawing on a powerful technical education and years of practical experience, coupled with an innate sense of business acumen, Rick Sievert presents this treatise on Total Productive Facilities Management.

Since one must admit that our times do not sanction the business-as-usual methods of the last decade, facilities must be evaluated in terms of their purpose, ability to serve customers, probable and potential uses, and expected life span.

In this book, *Total Productive Facilities Management,* Mr. Sievert offers a logical and methodical approach for rationalizing the relationships among costs, value, current price, and return on investment in terms comprehensible to both neophyte and seasoned facility professional. Here is a book that is a guide to doing it right the first time.

Among the topics discussed is benchmarking. Here, Mr. Sievert drives the reader to ask: Where is the facility now? What is possible? And, how far do we wish to go with the facility? A primary requirement for this portion of the book is an understanding that reality is an acquired taste. Readers must be willing to be honest with themselves and their stakeholders in the assessment of their facility and its future possibilities.

Among the many issues that arise are specific evaluations for performance. The book serves as a prompt for questions such as:

Does the facility do what it was intended to do? Does it comply with existing standards for its use? As Mr. Sievert guides readers through the methodology to the answers to these and similar questions, one can see how future performance goals can become reality.

Additionally, *Total Productive Facilities Management* demonstrates the current reality of regulatory compliance in facility management. It not only discusses the value of compliance to facility stakeholders, but also the costs and implications of noncompliance. This discussion demonstrates the necessity of regulatory and administrative compliance with such agencies as OSHA and EPA or local fire marshals and insurance companies. Mr. Sievert illustrates how the associated costs of these compliance issues should be factored into the overall facility cost equation and made part of the comprehensive facility management plan.

Further down the road to Total Productive Facility Management are project improvements. Mr. Sievert presents sound guidance for managing, as well as thinking about, projects. He guides the reader along the path to learning about what drives a project management process. One becomes aware of how specific, short-term project goals should mesh with the long-term purpose of the facility. This discussion includes the many aspects of a project life cycle—what will be attained, how it will be attained, and when. This particular part of this book illustrates the difference between continual facilities asset care or maintenance activities, as compared with discrete limits—and their respective outcomes and benefits.

The common thread in TPFM is *value* as applied to facility management. Mr. Sievert's methodology illustrates how one builds value through the practice of value engineering and other types of evaluation. One can learn through this book how to organize and prioritize thinking and actions to derive the maximum benefit from facility assets. Answers to such questions as: "What do we want in a facility?," "What is the current price?," and "What is it worth?" are the payback. Mr. Sievert shows how cost management ensures that facility managers or owners get what they specified at a reasonable cost.

Finally, Mr. Sievert describes how the essence of Total Productive Facility Management hinges on the necessity of good relationships—the crux of the issue is communications. He describes the functional and working relationships of the project team, contractors, facility planners, and the management teams and how their level of communication determines project success and profitable facilities.

For those who would get more from their facilities, Mr. Sievert's book is both a how-to guide and a reference work incorporating the best and most practical aspects of value engineering. It is also a road map to customer satisfaction.

Charles M. Boyles, C.P.E.
Editor-in Chief
Plant Services Magazine

Acknowledgments

The completion of this book was a team effort. It is a compilation of the author's experience, combined with insight, skills, and knowledge shared by others. The publications of industry professional associations, such as AACE International, the American Society of Heating, Refrigerating and Air Conditioning Engineers, the American Institute of Architects, the Association for Facilities Engineers, the Construction Specifications Institute, the Institute of Industrial Engineers, the International Facility Management Association, the Project Management Institute, and SAVE International, served as valuable references.

The author thanks Ed Hartmann, President, International TPM Institute, Inc., for his suggestion to develop the book approximately four years ago. Special acknowledgments are due to Chuck Boyles, Editor in Chief of *Plant Services* magazine (Putnam Publishing Company), for providing encouragement and helpful editorial comments concerning early drafts of the text; to Ed Deloria, Manager of Design and Construction for Ameritech Services Company, for contributing many of the engineering economics examples in this book; and Dimitri Karastamatis, Manager of Engineering Services at the *Chicago Tribune*, for sharing his perspectives on organizational design of facilities operations.

Major contributions to the material found in the maintenance management examples came from Ralph Grimse, Director of Support Operations for World Color, Chicago Division; and Larry Wolfgram, Plant Engineering Manager for Baxter Healthcare Corporation. Special thanks go to Bill Shull, Manager of the Operations Facilities Engineering Services Section for Fermi National Accelerator Laboratory; and Mike Prena, Manager of Facilities Services for Motorola, for providing invaluable insights on facilities benchmarking issues. Bill Shull also provided special knowledge in the area of outsourcing facilities operations.

For helpful suggestions in the preparation of this book, the author and publisher are also grateful to Paul Bouschard, Department Manager, Facilities Engineering for Uarco Incorporated; Geri Gaj, Sector Manager, Real Estate for Motorola; Ed Kestin, Senior Real Estate Project Manager for W.W. Grainger, Inc.; Pete Kopala, Senior Project Manager for Amoco Corporation; Rick Kriva, Director, Real Estate for Facilities Development and Environmental at Motorola; Mike Lamberty, Senior Facilities Engineer for Motorola; Dan Vanden Brink, Manager of Engineering Services, 3Com; Vince Serritella, Building Services Manager for Motorola; and Jim Warkentien, Sector Manager for Facilities Design and Construction at Motorola.

Harold W. Conner, Professor Emeritus, College of Architecture, University of Oklahoma; Professor James D. Stevens, Department of Civil Engineering, University of Kentucky; Philip C. Nunn, Retired Academic Dean, Master of Project Management Program, DeVry Institute of Technology; Professor Raymond J. Krizek and Professor Ahmad Hadavi, Department of Civil Engineering, Northwestern University; Professor Donald N. Frey, Professor Sanjay Mehrotra, and Professor Charles W. N. Thompson, Department of Industrial Engineering at Northwestern University; and Professor Robert Dewar and Professor Marvin L. Manheim at Northwestern University's Kellogg Graduate School of Management.

There are many colleagues at work who also provided support while this book was being developed. Bill Brandt, Scott Breneman, Pete Nomellini, Joe Scaccianoce, and Chris Jones stand out in particular.

Mary Greene, manager of the Reference Book Department at R.S. Means Company, Inc., edited the final manuscript and provided the gentle prodding necessary to bring this completed text to fruition. Phi Waier, principal engineer at R.S. Means, reviewed the manuscript and provided helpful comments on the technical content of the book.

My wife, Jane, also deserves recognition. She critiqued many rough drafts of the text. I owe more than can ever be repaid to her and my children—Sarah, Mark, and Matthew—who spent countless hours without me during evenings and weekends while I was struggling to complete this project. Ultimately, my gratitude is to God for all good gifts and for all those whom He has placed in my life.

Richard W. Sievert, Jr.

Part I

Evaluating Facilities Performance and Setting Goals

Chapter 1

Where Do You Stand Today?

We are entering an era in which the traditional ways of doing business will be increasingly challenged. Advances in computer and communications technologies, combined with transportation improvements and demographic and regulatory trends, are causing markets to shift, forcing businesses and individuals to cope with increasing competition. In many industries, fierce competition is driving down the rate of return on invested capital, resulting in the closing of facilities and downsizing of businesses.

While we cannot be sure of the coming economic trends, we do know we can no longer afford to operate our businesses as we did during more prosperous times. Businesses today are demanding that individuals do more; do it better; and do it with less time, money and personnel resources. We are all searching for a competitive advantage, and there is a heightened sense of the importance of improving efficiency in all areas.

For many organizations, buildings represent an underutilized asset. Facilities clearly impact a company's ability to produce its product, compete with others, attract customers, and attract and retain employees. Since many businesses have a major capital investment in facility resources, high asset utilization and capability are critical. As businesses audit their facility costs, they may find a sizable amount of money spent with little management control. Management needs better information and methods to evaluate the performance of resources allocated to facilities and to identify opportunities for improvement. Owners and managers of facilities must periodically and systematically analyze internal facility performance data, and compare that data with the organization's own standards and industry benchmarks. Many of today's leading companies are taking advantage of proven new computer and communications technologies, together with modern management techniques, to reduce costs, increase operational efficiencies, and deliver higher quality services.

A systematic method is needed for auditing your own facility's performance in these areas and implementing required changes.

Managed carefully, facilities can be used to help businesses maximize return on investment and establish a long-term competitive advantage in the marketplace. Businesses need to squeeze that last ounce of productivity out of their facilities for the lowest cost. Yet this task is difficult because economic resources for development, operation, and maintenance of facilities are scarce in most organizations. We need to change the way we think about and manage facility assets; the "same old way" is no longer good enough.

Total Productive Facilities Management (TPFM)

TPFM views facility management as a total problem-solving methodology. It is a concept that incorporates new management strategies and techniques to aid in facility planning, decision-making, and control of the multitude of activities involved in gaining a competitive advantage and getting the most return on investment from facilities, while supporting the organization's mission. TPFM draws from some of the best, proven strategies such as Quality Management and Benchmarking, Value Engineering, Life Cycle Costing, and the best project management techniques, as well as Total Productive Maintenance and team building. TPFM bridges the gaps between these management techniques and maximizes their effectiveness in identifying and implementing the dramatic changes needed to obtain a competitive advantage. This book explores these techniques, and others, as they fit into a comprehensive approach that maximizes your opportunities to get the best return on investment, while supporting customer requirements and the goals of the overall organization.

TPFM supports modern management precepts of constantly examining products, practices, and services from your customers' perspective, use of cross-disciplinary team problem-solving to reach consensus decisions, and continuous improvement. Customer attitude surveys, process flow charting, and benchmarking are some of the tools of TPFM. Systematic application of these techniques will enable facility owners and managers to understand and satisfy customer requirements, achieve quality improvement, enhance productivity, identify and remove unnecessary costs, decrease cycle times, and improve overall competitiveness and return on facility assets.

TPFM works in a team environment, focusing on people working together and utilizing data to improve the systems, processes, and methods by which we monitor and evaluate facilities; plan and program facilities; and design, build, operate, and maintain facilities. A multidisciplinary approach is an important part of TPFM. Problems that appear to be quite simple can become profoundly complicated when viewed from the perspectives of various disciplines. A solution from one perspective can create a host of problems from another perspective. TPFM calls upon the best of management techniques and teaming of departments to help owners, facility managers and operators to build, remodel, rearrange, and improve their facilities.

Definition of Facility Management

Facilities must be managed as an integrated system. The International Facility Management Association (IFMA) defines facility management as: "the practice of coordinating the physical workplace with the people and work of the organization."

Facility management integrates the principles of business administration, architecture, engineering, and building construction, and the behavioral sciences (e.g., psychology, sociology). Facility management involves the development of:

- Corporate policies
- Long-range forecasts
- Real estate strategies
- Space plans
- On-going equipment and furniture inventories
- Project planning and programming
- Project management
- Design and construction for renovation and alterations
- Energy management
- Workplace health and safety
- Operation and maintenance

Facility management may also involve providing support services for lease management, security, voice and data systems, food and vending, mail, company vehicles, receiving and shipping, purchasing, and office management functions. Essentially, facilities management encompasses all of the tasks required to make a facility function in accordance with an organization's strategic plan and occupant needs. Facility managers plan, organize, direct, and control facility assets and services to support the objectives of the business as a whole. You must think of your facility as a strategic business resource.

Investing in new facilities, equipment, and personnel to improve performance can, in some cases, actually make things worse. Managers need accurate and sufficient data to identify and understand problems before making changes. If management has not addressed the root causes of productivity-related problems, investing in ill-advised changes can cause productivity to decrease even further.

Each facility management activity is part of an overall process. The focus must be on *how* the work gets done (the methods) instead of simply *what* is done (the results). You can only achieve breakthrough improvements by understanding and changing the process. Complete and accurate data is necessary to work on process improvement.

The Changing Facility Management Profession

If you think you have no reason to change or consider alternative approaches to managing your facilities, think again. We all must realize that survival in our dynamic and competitive marketplace requires that we change the way we do business and that we relate more closely to our customers. Constant, rapid change is the norm, and the old adage, "If it ain't broke, don't fix it" is obsolete.

Change is necessary in facility management groups that wish to continue their corporate viability. Facility managers must get the most from every asset within their control or they will be replaced by somebody who can. You must demonstrate in financial terms that management can understand and appreciate that you are an asset to the business worth maintaining. You must keep pace with changing technology and management techniques and continue to learn. You must be a leader and chart a course of action that your customers will value and others will follow willingly.

Facility Management in the Past

Facility management was formerly regarded as strictly an overhead or maintenance function. Facility managers oversaw operation of the boilers, handled equipment failures, changed light bulbs, moved furniture, and swept and vacuumed the floors. Facilities were considered cost centers rather than assets to be managed carefully. As such, facility management was seldom regarded as a strategic planning function, critical to an organization's competitive advantage. The building systems were less complex and easier to maintain. Energy was cheap. Building codes were not as stringent. Occupational Safety & Health Administration (OSHA) and Environmental Protection Agency (EPA) requirements were nonexistent. We were less sensitive to the impact of the building environment on human needs. Worker comfort was not important because it was thought that people could adjust. Facilities were erected to "contain work" rather than "support work."

Facility Management Today

Today, facilities are viewed as strategic business resources. The position of the facility manager has been elevated to a higher level in the business organization. The facility manager is considered an asset manager, responsible for supporting the overall organizational objectives.

The new attitude calls for a company to adjust the work space to fit occupants' needs — not the reverse. In our competitive and uncertain environment, with emphasis on quality, cost reduction and profitability, owners are concerned with getting a better return from investments already made by the organization in facilities and equipment. The emphasis is on maintaining, repairing, renewing, and managing existing facilities, rather than building a lot of new facilities.

Many facility managers will tell you that they lack sufficient time and funds to perform their many duties properly. They may feel compelled to buy time by deferring some actions that really should be done immediately. For example, facility managers may decide to make certain repairs to existing systems, knowing that the optimal investment would be in a new facility. Moreover, facility managers are often overpowered by interrelated and conflicting objectives, and pressured to improve working conditions within buildings while reducing capital and operating costs — all at the same time. Much of their time is spent responding to emergencies and building occupants' varied requests that do not fall under the charter of other departments. Consequently, it can be very difficult to schedule adequate time for strategic planning and continuous improvement projects.

The government continues to enact stricter air quality laws mandating requirements for designing, operating, and maintaining building mechanical and process systems. The goal is to achieve both a healthy environment and energy savings. Improving the air quality within a facility may require the installation of additional exhaust to remove contaminated air, and makeup air equipment to increase the amount of outside air supplied to occupied spaces. The energy cost probably will increase because the additional outside air has to be tempered with heating or cooling to meet indoor temperature and humidity requirements. It may also be necessary to install pollution control equipment to comply with government requirements. It may seem as if the cost for these improvements contradicts the objective to reduce capital and operating (including energy) costs. On the other hand, the cost of improving ventilation, exhaust, and filtration systems is a worthy investment for an organization concerned about reducing employee sick time and potentially increasing productivity. Aside from these benefits, there is the necessity to avoid potential fines, costly lawsuits, and even possible prison sentences for willful and wanton noncompliance with government-imposed air quality mandates.

Since manufacturers no longer produce chlorofluorocarbons (CFCs), facility managers must choose between keeping existing cooling equipment that uses CFCs or purchasing new equipment. If they keep

Seven Tensions of Facility Management
1. First cost capital budget vs. life cycle cost
2. Occupant health, comfort & productivity vs. operating cost
3. Indoor air quality vs. energy
4. People needs vs. profit objectives
5. Available time vs. schedule
6. Delivery vs. price
7. Family vs. work

the old equipment, they must either contain and maintain it, paying the exorbitant cost for purchasing stockpiled CFC refrigerants when needed, or converting it so that an alternative refrigerant can be used.

There are rumors that some companies have assigned an individual who will accept the blame if charges are pressed against the company for sacrificing public health, safety, and welfare in order to avoid costs associated with installation and operation of more expensive pollution and safety control technologies required by law. The fact is that if you cut the staff and programs that support the environment and safety requirements, you may be held legally responsible.

Another issue faced by facility managers is the need to provide for a growing business. On the one hand, you do not want to commit to a project that will put the organization in too much debt, but on the other, you need to expand in order to increase your market share and earning potential. If you spend too much on your facility, you may have to fund this expense by either charging more for your product and possibly losing some of your market share, or selling more than is projected in an economy where there may be limited demand for your product. How much more money should be invested in a new facility to provide flexibility for unknown future requirements? The answer to this question will depend on the goals of the particular organization, its business plan, and environment.

Most owners and developers of facilities prefer to include a certain degree of flexibility in their buildings. However, the cost to provide flexible space may be greater in the beginning phase of a building's life cycle because the systems will most likely be designed with more capacity than initally required by the space users. The design of flexible space may entail floor and roof structures that can support the weight of heavier equipment, increased floor-to-ceiling heights, larger floor areas, wider column spacing to accommodate changing circulation patterns and rearrangement of production equipment, additional cooling and air handling system capacity, more temperature control zones, additional plumbing and heating, and a larger and more flexible electrical service. In high technology research and development and manufacturing environments where change is constant, the cost of flexible support systems and space layouts is often justifiable from a building life cycle cost standpoint. Also, in dynamic and competitive industries such as electronics and biotechnology, quick and constant change of space is essential to lower manufacturing costs, improve quality, and accommodate new products and processes. If you lease a facility, the additional cost of flexible building systems may be more difficult to cost justify. Techniques such as life cycle costing and value engineering are very useful in helping to evaluate investment decisions of this type.

Staying Competitive While Controlling Costs

Competitive mandates require continuous innovation and improvement. There is also a need to deliver facilities faster because time is money, and product life cycles tend to be shorter in our rapidly changing marketplace. However, you do not want to pay a premium to reduce the lead time for delivering facility projects.

Code requirements, production, sales, and occupant needs all have to be addressed. All are competing for your organization's limited funds. The trade-offs must be considered and conflicts resolved between performance mandates in a timely manner. Priorities must be set based on the functions of the facility or building service.

Owners and facilities managers, facilities engineers, and maintenance managers must plan and implement facility strategies to ensure the success of their organizations. The scope of services traditionally provided by these disciplines is clearly affected by rapidly changing business conditions. It is imperative to adopt a philosophy and a culture attuned to the needs of the changing corporate environment.

Process improvements in such functions as corporate facility management, facilities engineering, and maintenance may have been among the last to be considered, but they can no longer be ignored. To contribute to corporate competitiveness and profitability, facility assets must be properly positioned. Some businesses may have to learn how to survive on smaller orders and tighter profit margins. This means as a facility manager, you may have to learn to live with less.

Generally, global competition, declining markets, and technology developments have resulted in the need for fewer permanent employees and less space. We are striving to satisfy our customers by producing higher quality products and services, faster, less expensively, and with fewer resources. Now and in the future, cost control is an absolute necessity.

Since facility costs are high for many companies, attention is being turned to accountability for expenditures. You probably are going to have to cost justify more things than you have ever before, including your own value to the organization. Mounting pressures to reduce facility-related costs and liabilities have resulted in plant closings and downsizing of facility department staffs, development of new management methods, and the use of specialist consultants that challenge traditional facility management practices.

As organizations focus their resources on conducting their revenue-producing core business, there is a growing trend toward outsourcing, and the facility manager's traditional scope of services is changing. The old approach of having company employees carry out janitorial, groundskeeping, security, food service, reprographic, maintenance, and building management services is being challenged.

Computer and Communication Technologies

More and more companies are looking for new ways to take advantage of developments in computer and communication technologies and emerging management strategies to reduce work space and associated costs, while making their workplaces more productive. Some of the new tools include:

Telecommuting: An arrangement that gives employees flexibility to work out of their home or other locations without having to commute to their employer's office every day. Telecommuters are linked to their organizations over a telephone network through the use of modems, computers, and fax machines.

Video Conferences: A meeting tool that allows groups of people situated in different locations to communicate with each other using television monitors, screens, or computer terminals. Videoconferencing saves on costs associated with long-distance travel and lodging, and allows people to attend training or other conferences without physically being there.

Virtual Offices: An office that travels with the person. Personnel are equipped with cellular phones, battery-powered laptop computers, modems, and portable fax machines, and can function from virtually any location.

Hoteling: An arrangement where employees who spend most of their time outside the main office call to reserve a temporary work space within their home office for a requested period of time. This gives an organization the advantage of not having to provide permanent work space for personnel, such as field sales representatives, accountants, and construction site superintendents.

Team Spaces: Space provided for a multi-disciplined group of people to work together in the same location throughout the project duration — especially useful for extended projects.

The Facility Manager's Challenge

Facility professionals must develop a keen business sense and well-honed leadership, communications, team-building, and problem-solving skills. Since many senior executives neither understand nor are interested in learning more about building operations and maintenance, it is necessary to translate facility requirements into terms they can understand, such as *capital budgets*, *return on assets*, *payback period*, *return on investment*, *life cycle costs* and *strategic business planning*, as well as legal and productivity issues, including life safety and exposure to liability. It is critical to demonstrate how your facility management expertise and services support financial, production, distribution, marketing, environmental compliance, quality, and other strategic plans of your business. The facility department's ability to serve as a financial asset and a profit center must be emphasized. If the facility management department is viewed as a non-revenue producing asset or a cost center that is unable to compete with outside services, it will be outsourced.

Facility management organizations are reinventing and remaking themselves. Facility managers are being forced to do more and more with less and less. Productivity increases in support service work are imperative. There is a desperate need for systematic and innovative problem-solving and decision-making techniques. Total Productive Facilities Management addresses that need with a comprehensive approach that integrates value engineering, benchmarking, cost management, and team-building with operations and project management.

Evaluating Your Management Approach

The following two exercises are designed to help you review your current management practices. They are followed by a list of action items to help you improve your approach.

Exercise 1

List the significant changes to your physical facilities and management practices that you have implemented within the past two years to make your organization more competitive and to improve overall quality of facility management. As you begin to list these changes, think about adjustments to operating policies and procedures, organizational structure, all management techniques, and attitudes (such as acceptance of telecommuting and flexible work hours, and interest in employee comforts), in addition to the physical facility changes.

Exercise 2

Identify major conflicting goals you face, and your technique for dealing with the conflicts.

Planning and Organizing a Program To Achieve Continuous Improvement

Is your organization improving with respect to facility management? Improvement alone is not enough. You must improve at a faster rate than your competitors if you and your business are going to survive and prosper. Following are some action items for improvement.

- **Make a commitment to set aside time on a regular basis for improvement work.** The same types of problems tend to occur over and over again in organizations. Too often managers only put out fires, addressing symptoms rather than the root cause of problems. Probably one of the biggest problems in an organization is making time for improvement. For that reason alone, many organizations that start improvement programs fail.
- **Make the most out of every asset you have, but continually search for more productive ways to use your abilities and facility resources to increase your company's profitability.** To do this, you must collect data and analyze facility functions and resources as they relate to the business climate, then systematically apply

team problem-solving techniques. Look for common patterns in problems, and use process improvement techniques to identify and solve the problems.

- **Understand which activities your customers value and what level of performance they expect.** Question every practice. Ask yourself and your customers if it is necessary. "Why is this being done? Can it be simplified, combined, reduced, or eliminated?" "What am I doing right *and* what am I doing wrong?" When you have a problem, ask why until you get to the root cause. Asking the right questions tends to lead to the one fundamental cause of many problems.

- **Always examine how and why you do things.** If you are unable to justify how a specific activity or function reduces costs or adds value to your company, then you had better find something more productive to do. Concentrate available resources and services on helping your customers be more productive, more competitive, and more profitable, which will ultimately increase customer and user satisfaction.

- **Find ways to improve existing conditions and promote changes needed to accommodate new requirements.** The improved performance is directed toward satisfying cross-functional goals, such as "6 sigma quality" (3.4 defects per million opportunities), cost reduction, and cycle time reduction.

- **Make plans that support your organization's mission.** Like corporate management, the facility management organization also needs a clearly defined statement describing succinctly what the organization does and for whom. Sometimes called a "charter" or "purpose," the mission has to create a sense of common purpose. There is an old engineers' saying: "When you're up to your neck in alligators, it's hard to remember that the mission was to drain the swamp." Being able to refer to a charter at such times helps keep people on track.

  The facility management mission statement should define the facilities link to the overall organization's strategy; the scope of the facility management; and its strategic vision, its objectives, and its responsibility and authority. You must challenge what needs to be challenged in order to achieve business objectives.

- **Work with your customers and suppliers as a team to produce desired results.** Create multi-disciplinary teams, determine organizational strategies, map processes, establish effectiveness measures and benchmarking targets, and set up project management systems to implement organizational strategies. The following section provides guidance on analyzing your customers' requirements.

Exercise 3

What key measurements are used by senior management to rate or evaluate the performance of your facility management? (These might

12

include financial performance; safety; customer complaints; and the ability to deliver projects within time, cost, and quality constraints.)

Know Your Customers

Everyone in an organization has internal and external customers. Internal customers are people within your organization who receive the direct output of your service. Your internal customer is a business associate, department, or division whose ability to do their job depends on how well you do your job. An external customer is the individual or group who purchases or uses your final product or service. Your customer may be a department affected by the facility support service, who in turn may be responding to another internal customer, who is ultimately responding to an external customer.

Lots of people talk about focusing on the customer, but do not know how to do it. Your customers may know what they want, although they do not articulate it. It is your responsibility to create opportunities for serious dialogue with your customers. You have to know how to listen to them and must ask the right questions in order to understand their goals, needs, and expectations. You need to know what your customers value and what they do not value.

One of the best ways to obtain useful information for improving your products and services is to conduct face-to-face interviews with individual customers or focus groups. In order to obtain more information about your customers' needs, concerns and perceptions, ask them open-ended questions, demonstrating your sincere interest. Do not ask only "yes" or "no" questions. Examples of appropriate open-ended questions include: What do you like about our service? What don't you like about our service? What do you consider a quality installation? What do you consider a good design? Ask not only what is important to your customers, but *how* important each item is.

Request the truth no matter how much it hurts. You need honest and complete customer feedback in order to improve your customer service. Use criticism to make necessary changes in a productive way. Your customers' reaction to features and characteristics of your product or service is necessary for you to learn how to satisfy their needs and serve them better.

Quality is measured by conformance to customer requirements. Your perception of quality may be different than your customers' idea of quality. The important thing is to refrain from putting your own spin on recording customer needs and perceptions. Record the information in your customers' own words.

One cannot assume that a customer is fully satisfied with your service or product just because you have not heard anything to the contrary. Customers have a tendency to avoid people with whom they are unhappy, so you need to solicit their feedback continuously to uncover opportunities for improvement.

You must create a system for regular recording of customer needs. Your business survival depends on supplying the services your customers desire *when* they desire them, and *at a price they are willing and able to pay.*

Customer Attitude Survey

Conducting a customer needs analysis within the facility management department should aid in identifying specific opportunities for improvement. Methods commonly used include surveys, questionnaires, comment cards, review of records and documents, interviews, and observations of how current space configurations and building systems (e.g., HVAC, lighting, power and telecommunications systems) affect productivity.

To meet or exceed customers' expectations, all of their needs must be defined and addressed. Also, it is necessary to distinguish needs from wants, because most customers want more than they can afford. A customer attitude survey is a powerful technique for acquiring, clarifying, and recording customer and user information. Knowing your strengths and weaknesses as perceived by your customers will provide you with the insights needed to identify opportunities for improvement that will significantly strengthen your position. A customer attitude survey involves facility representatives and their key customers, including representatives from senior management, production, engineering, human resources, legal, and the accounting department.

The customer attitude survey should focus on:
- Facility resources, services, and processes that support business processes
- Strategies and priorities that are important to key customers of the facilities department

The customer attitude survey can begin with a meeting in which the facilitator asks the participants to list the features and characteristics they feel are important to the success of the product, process, or service from their own perspective. Of those items, customers can identify which ones they feel are currently being handled well. All features and characteristics are recorded on a form similar to Figure 1.1, exactly as stated. Now that's really capturing the voice of your customer! [The method summarized in this chapter and in Figures 1.1–1.4 adapted from *Value Analysis Systems Training* by Robert Null.]

After all features and characteristics are listed, the facilitator gives each participant a set of numbered cards (1–10, excluding the number 5). These cards are used to vote and rank the importance of the features and characteristics under consideration. The facilitator asks the participants to identify which feature or characteristic listed is the most important by voting on its significance (1 = least, 10 = most). Everyone votes at the same time, and the average score of the team is recorded. The reason for omitting the number 5 is to prevent participants from being fence-sitters and not taking a pro or con position. The practice of having everyone vote at the same time helps to neutralize any one dominant personality whose strong influence or position of power may prevent others from expressing their individual opinions.

The other features or characteristics are evaluated in terms of importance by comparing their ranking with the one voted most important. This procedure can also be used to identify and rank faults and complaints (opportunities for improvement). (See Figures 1.3 and 1.4, the "Faults and Complaints Rating Sheet," a blank and a filled-in sample.)

The next step is to publish the results from this survey and have the team examine them to identify needed improvements. In addition to its value as a written document for making improvements, the *process* of conducting a customer attitude survey also shows customers you really care about serving them.

Figures 1.1 and 1.2 are sample Customer Attitude Survey Rating Sheets. The example in Figure 1.2 involves an industrial research and development facility. Some appropriate "positive feedback" items have been listed and ranked by a team of six facilities customers. This group is participating in a post-project evaluation session. Bill is the project sponsor, responsible for investing the money in the project. Steve represents operations and maintenance. Ramon is the project manager — responsible for cost, time, and quality of the project. Chris is the Director of Facilities. Jim is the Engineering Manager, and Claudia is the Accounting representative.

Figures 1.3 and 1.4 are additional ratings sheets, this time listing and ranking negative feedback items from a completed project.

Conducting the survey and collecting the information (positive and negative) gives the Service provider (the facilities department) the opportunity to identify the changes they need to make to improve their contribution to overall organizational profitability in the future.

Know Your Environment

A thorough diagnosis and analysis of your current environment is a necessary step towards meeting your goals. Your assumptions about the environment — which includes the status of your organization and its structure, the marketplace, customers, competitors, and technology — must be compared realistically with your core competencies and organization's mission. An environmental analysis is an essential component of the facilities planning process. We must understand our strengths and weaknesses in relation to external opportunities and threats, so that we can take the necessary action in a timely way.

SWOT is an acronym for an organization's *internal* Strengths and Weaknesses and its *external* Opportunities and Threats. (This term was used by Arthur A. Thompson, Jr. and A.J. Strickland III in the

I. FEATURES & CHARACTERISTICS RATING SHEET

PAGE ____ of ____
DATE ____
RATED BY ____

PRODUCT, PROCESS OR SERVICE: ____

CUSTOMER: ____

#	POSITIVE FEEDBACK (10 = most important / 1 = least important)	IMPORTANCE 1 - 10 (average/spread)								

Note: In some circumstances, after discussing the expressed importance of a particular feature or characteristic, the team may decide to call for a re-vote. A second column can be added on for a re-vote, if necessary.

Figure 1.1

16

1. FEATURES & CHARACTERISTICS RATING SHEET

PAGE ____ of ____

DATE ____
RATED BY ____

PRODUCT, PROCESS OR SERVICE: Division Facilities Engineering Department

CUSTOMER: General Manager

#	POSITIVE FEEDBACK (10 = most important / 1 = least important)	IMPORTANCE 1 - 10 (average)R(spread)	Bill		Steve		Ramon		Chris		Jim		Claudia	
			#1	#2	#1	#2	#1	#2	#1	#2	#1	#2	#1	#2
1	Internal knowledge of the construction process enabled the team to make decisions that minimize cost and delivery time.	9R2	9		9		8		10		9		8	
2	Internal customers benefitted from the team's ability to define the project requirements and scope of work.	6R2	8	8	7	7	4	6	4	6	6	6	6	6
3	The project was kept within the scope, cost, time and performance requirements	10R0	10		10		10		10		10		10	
4	The department is looking out for the interests of the corporation more than outside consultants/contractors. Project team preserved corporate usage, while meeting all company policies, standards, and project objectives.	10R2	10		8		9		9		9		10	
5	The team's close proximity and availability for customers for enhanced communication.	9R2	8		7		7		8		8		8	
6	The facilities department's past history and reputation for satisfying customer requirements enhanced its ability to complete this project successfully.	8R3	8		7		6		9		8		8	

Note: In some circumstances, after discussing the expressed importance of a particular feature or characteristic, the team may decide to call for a re-vote. A second column has been added on for a re-vote, if necessary.

"R" indicates the range between high and low vote.

Figure 1.2

17

Figure 1.3

II. FAULTS & COMPLAINTS RATING SHEET

PAGE_____of_____
DATE_______________
RATED BY_____________

PRODUCT, PROCESS OR SERVICE:

CUSTOMER: ________________________________

#	NEGATIVE FEEDBACK (1 = most important / 10 = least important)	IMPORTANCE 1 - 10						

Note: In some circumstances, after discussing the expressed importance of a particular feature or characteristic, the team may decide to call for a re-vote. A second column can be added on for a re-vote, if necessary.

II. FAULTS & COMPLAINTS RATING SHEET

PRODUCT, PROCESS OR SERVICE: _____

CUSTOMER: _____

PAGE __ DATE __ RATED BY __

#	NEGATIVE FEEDBACK (1 = least problem / 10 = worst problem)	IMPORTANCE 1 - 10 (average)R(spread)	Bill #1	Bill #2	Steve #1	Steve #2	Ramon #1	Ramon #2	Chris #1	Chris #2	Jim #1	Jim #2	Claudia #1	Claudia #2
1	Operations viewed engineering department as competition for funding and allocation of scarce resources.	8R4	10		8		6		8		7		7	
2	Internal client's requirements differed from corporate guidelines that the engineering department had to follow	9R3	9		8		6		7		9		9	
3	Personality conflicts occurred between individuals from different departments who were forced to continuously work together.	4R8	9		3		1		2		4		4	
4	Routine operations were somewhat disrupted due to the inability to plan projects with the building supervisor and occupants.	8R3	6		8		7		9		8		8	
5	All project stakeholders were not involved in the project definition and development of the scope of work.	10R3	10		8		10		7		10		9	

Note: In some circumstances, after discussing the expressed importance of a particular feature or characteristic, the team may decide to call for a re-vote. A second column has been added on for a re-vote, if necessary.

"R" indicates the range between high and low vote.

Figure 1.4

19

book, *Strategy Formulation and Implementation. Tasks of the General Manager*, published by Business Publications, Inc., 1986.) To develop a strategy that is the best match between opportunities and a firm's facility resources, it is necessary to appraise the strengths and weaknesses of your internal resources with respect to the external market opportunities and threats. The SWOT analysis involves development of a set of four lists: one for the facility operation's strengths, one for its weaknesses, one for its opportunities, and one for its threats. Strengths are those internal resources or conditions that might help you achieve your goals. Weaknesses are internal obstacles that might prevent you from achieving your goals. Opportunities and threats are external conditions that impact your internal strategies and goals.

Based on a SWOT analysis, owners and managers of facilities develop strategies for dealing with environmental and competitive pressures. The SWOT analysis can help identify needed changes in an organization's mission, policies and practices, as well as core competencies that must be developed and acquired to meet your organization's objectives. Cost avoidance programs, energy conservation measures, downsizing and outsourcing, indoor air quality surveys, Americans With Disabilities Act compliance audits, and ergonomic programs are the kinds of efforts that may result from a facility manager's SWOT analysis.

Figure 1.5 is a sample SWOT analysis, listing a hypothetical facilities organization's strengths, weaknesses, opportunities, and threats. Figure 1.6 is a blank form that you may wish to use to perform your own department analysis as an exercise.

Summary

Recognizing the need for constant improvement in productivity, customer satisfaction, and meeting the overall goals of the organization is the starting point for implementing a plan for TPFM. The next step is assessing where your facilities department stands today, using tools like the SWOT Analysis, the positive and negative feedback project rating sheets, and self-evaluation exercises.

The next chapter reviews some of the standards facilities should be meeting, as a part of the criteria for evaluating their performance. The chapter describes the types of evaluations that should be conducted for particular purposes, such as pre-acquisition, post-occupancy, and for requirements programming.

SWOT ANALYSIS

INTERNAL

Strengths	Weaknesses
1 Familiarity with current facilities and operations	1. Inability to identify market opportunities for facility management.
2. Multi-disciplined staff of real estate and facility management personnel.	2 Lack of experience in new fields of opportunity
3. Low first-cost provider. (No profit markup or internal staff.)	3. Lack of a systematic approach to identifying and implementing required changes.
4	4 Fear of job security.
5	5. Lack of strategic partners.
6	6. Inefficient communications management.
7	7. Lack of firm grip on all existing assets and facility resources.
8	8. Lack of time.
9	9 Limited financial resources
10	10 Operating in environment of constant change can lead to wasted resources and communication problems

EXTERNAL

Opportunities	Threats
1. Implement projects that satisfy new corporate strategies	1. Increasing global competition in traditional lines of business
2 Downsize space standards to improve space utilization	2. Less predictable business growth.
3. Form strategic alliances with other companies	3 Increasing emphasis on cost-cutting measures.
4. Conduct value engineering studies and workshops.	4. General decline in facility square footage requirements.
5 Modernize and change work spaces to accommodate new technology equipment standards and management concepts	5. More stringent governmental regulations and insurance requirements
6. Conduct code compliance studies.	6. Redundant staff due to increased acquisition, merger and divestiture activity.
7 Provide time to develop internal skills and rely less on outside resources	7 People and traditional jobs replaced by technology.
8. Conduct financial reserve studies.	8 Acceptance of outsourcing as a viable way to manage facilities.
9. More time to evaluate and improve performance of existing assets	9. Senior managers' perception that facility management is a non-essential (core) service.
10. More time for maintenance that will extend equipment life and value.	10. Need to reduce cycle times.
11	11 Fewer traditional projects to manage

Figure 1.5

SWOT ANALYSIS

INTERNAL

Strengths	Weaknesses
1.	1
2.	2.
3	3.
4.	4
5	5
6.	6
7	7
8.	8
9.	9.
10	10.

EXTERNAL

Strengths	Weaknesses
1.	1.
2	2.
3.	3.
4	4
5	5.
6.	6.
7	7
8.	8.
9.	9
10.	10.

Figure 1.6

Chapter 2

The Need for Facilities Performance Evaluations

The quality of facility managers' decisions is directly related to the quality of information available to them. Facilities evaluations provide decision-makers with information about the facility's total performance and whether the requirements of customers, users, and outside regulatory bodies are being met. The data used in these evaluations is accumulated through different vehicles such as the customer attitude survey, questionnaires, and observations, as covered in Chapter 1.

In this chapter, we will review types of facility evaluations and the circumstances under which they are used. These include pre-acquisition surveys, post-occupancy evaluations, and requirements programming. In addition to the organization's own functional needs, government and industry standards, such as building codes, and association recommendations, must be addressed.

Standards

Managing facilities for optimum performance also requires information about life cycle costs, productivity attributes, quality of the physical work environment, the facility's physical condition, and the functional adequacy of the buildings the organization uses or intends to use. This information is the basis for conducting systematic performance evaluations to determine conformance with existing specified requirements and standards, and for setting new facility performance requirements and standards.

Development of standards can help the manager determine the improvements needed and evaluate available alternatives. In prioritizing proposed improvements, it is worth noting the following facility expense ratios. In determining the impact of buildings on overall organizational effectiveness, these figures were reported in *Using Office Design to Increase Productivity* by Michael Brill, Stephen Margulis, Ellen Konar, and the Buffalo Organization for Social and Technological Innovation (BOSTI, 1985). Over the life of a newly constructed building, personnel salaries represent about 93% of the cost of owning and operating a structure. Construction, furniture and equipment represents 5%, and maintenance and operations 2%. The conclusion is that building improvements made to increase occupant productivity can result in significant savings to a business.

Code Compliance

Organizations have a moral and legal obligation to provide a safe environment for building occupants. Building codes and government regulations mandate the minimum requirements and standards for public health and safety. Businesses are not permitted to expose workers to unsafe conditions in the interest of saving time or money, using such approaches as decreasing ventilation air below code-required levels in order to reduce energy costs, or increasing productivity rates to unsafe levels.

OSHA and building code enforcement agencies hold owners and operators accountable for the safe and sanitary maintenance of their facilities. If a building inspector approved or overlooked a building defect or code violation, it does not mean that the manager is off the hook. As a result, some owners' attorneys now recommend periodic evaluations of occupied buildings to determine code compliance, and to guard against potential claims from people who may blame building mismanagement for their physical or mental health problems. Owners are being asked whether they did everything "reasonably possible" to prevent alleged harmful occurrences. States are prosecuting violators

More Humane Working Environments

There is an increasing emphasis today on providing more humane environments for people. Some organizations have developed their own voluntary standards which often exceed the mandatory, but sometimes antiquated, building codes and government regulations. In addition to indoor air quality issues, consideration should be given to thermal comfort; power distribution and illumination; telecommunication systems; and structural, architectural and interior design, acoustics, and ergonomic features as they relate to worker health, morale, and productivity.

The following organizations collect and publish data regarding facility operating conditions and productivity standards:

- **APPA: The Association of Higher Education Facilities Officers**, 1643 Prince Street, Alexandria, VA 22314-2818, (703) 684-1446
- **The American Society of Heating, Refrigerating and Air-Conditioning Engineers (ASHRAE)**, 1791 Tullie Circle NE, Atlanta, Georgia 30329-2305, (404) 636-8400
- **The Association for Facilities Engineers (AFE)**, 8180 Corporate Park Drive, Suite 305, Cincinnati, OH 45242, (513) 489-2473
- **American Society for Testing & Materials (ASTM)**, 1916 Race Street, Philadelphia, PA 19103-1187, (215) 299-5400

24

- **Illuminating Engineers Society (IES)**, 345 East 47th Street, New York, NY 10017-2377, (212) 705-7926
- **Building Owners and Managers Association (BOMA) International**, 1201 New York Avenue NW #300, Washington, DC 20005, (202) 408-2662
- **International Facility Management Association (IFMA)**, 1 E. Greenway Plaza, 11th Floor, Houston, TX 77046-0194, (713) 623-4362.

In addition, publications such as *Means Facilities Maintenance & Repair Cost Data* (R.S. Means Company, Inc.) offer productivity information with related costs for maintenance and repair tasks.

Types of Facility Evaluations

There are as many different types of facility evaluations as there are facility-related items to manage. The type of evaluation you select should be determined by the decisions that need to be made regarding each facility. Building evaluating can be broad-based, as in the example of a comprehensive pre-acquisition survey or post-occupancy evaluation of total building performance. Some evaluations are narrow in scope and may focus on specific performance attributes. Evaluations may focus on issues such as code compliance, indoor air quality, energy efficiency, handicapped accessibility, environmental hazards, maintenance and operating costs, user satisfaction, comfort and productivity, or space and workflow efficiency.

Pre-acquisition evaluations provide information that is needed to establish the condition, life expectancy, and functional adequacy of existing building systems. Financial reserve studies determine the funds required over a period of time to maintain building systems properly and to meet planned needs. Code compliance issues and performance standards should be considered in such a study, along with accessibility to disabled users (a civil rights issue) and environmental conditions, which may require investigations to determine the presence of potentially hazardous contaminants in the building environment.

Varying levels of effort are required for different types of facilities evaluations. Initially, preliminary walk-through surveys are conducted to make a cursory visual assessment of building systems and components. A walk-through survey may be a one-day tour of a facility to identify areas where energy is not used efficiently and to recommend further action. If required, more comprehensive diagnostic surveys and in-depth investigations would be made. These often involve using instrumentation and physical measurements, quantitative analyses, and preparation of detailed recommendations. A comprehensive diagnostic survey may include testing of heating, ventilating and air-conditioning systems using instruments for recording temperature, humidity, static pressure, fan speeds, air quantities, and power usage. Mechanical engineers may calculate thermal loads, prepare design solutions, and determine life-cycle costs and code compliance while considering energy-saving options.

Facilities Programming

Facilities programming is the process of evaluating and defining the requirements for a new or existing space to meet organizational objectives. Programming involves collecting and organizing qualitative and quantitative data about functional purposes of the space, tasks performed within the space, occupant requirements, production equipment and support systems, and environmental and economic criteria. Whether you are maintaining or changing the function of an existing space, or constructing a new space that is purchased or leased, it is important to properly program the space before making a financial commitment to the property.

It is necessary to define requirements in terms such as:

- Function and use of space
- Floor area
- Activity adjacencies and workflows
- Equipment
- Floor and roof loading
- Plumbing and drainage
- Power and lighting
- Security
- Life safety systems
- Telecommunications and data systems
- Temperature control, humidity, air cleanliness
- Ventilation
- Occupant activities and population
- Work area sizes and furniture
- Ceiling heights
- Acoustic features
- Flexibility to accommodate changing needs
- First cost, operating and maintenance costs
- Aesthetic features
- User amenities

The overall facility requirements program must support production, distribution, marketing, financial, quality, environmental compliance, and other strategic plans essential to the success of the organization. Departmental managers should be asked to complete questionnaires, charts, and forms indicating their needs and wants.

Programming is a separate and distinct pre-design service which is not typically part of the basic design contract services provided by architects and engineers. Initially, the facility requirements are defined in terms of functions or space usage by the owner or user of the space. The next level of this function is then turned over to a programming specialist, who helps translate the functional requirements into technical specifications that the space design must meet. Depending on the level of expertise, experience and availability of in-house personnel, the programming may be completed internally or with the help of external industry consultants.

The programmer assists the facility manager in framing project issues and documenting more specific criteria, addressing requirements with technical solutions. These are needed to evaluate alternatives and negotiate the final purchase or lease of a property, whether it is undeveloped land or an existing facility. The program requirements also become the basis for communicating the user's design criteria, and should be included in any subsequent contract for needed design and construction work between the owner and architect/engineer. Preparation of program requirements and documents identifying the scope of the project work should be completed before committing full capital funds to a project.

Managers should not evaluate prospective sites without first evaluating their present facilities. Before planning new facility projects, managers need to determine existing problems and inefficiencies in their current facilities and operations, so as not to recreate these problems in the new facility. Post-occupancy building evaluations uncover this kind of information. Building researchers classify the evaluations by technical, functional, and behavioral performance criteria.

Pre-Acquisition Evaluation

Figure 2.1 is a list of items typically addressed when evaluating a property prior to acquisition.

Pre-acquisition evaluations involve examination of a building and grounds prior to final purchase or lease of a property. A pre-acquisition evaluation is conducted to determine the general condition and functional adequacy of an existing site and building. It identifies any remedial repairs and upgrades which may be necessary, and includes preparation of budget cost estimates. The pre-acquisition survey is helpful in negotiating the final purchase or lease of the property. It is typically required by a lender for financing the property. Pre-acquisition surveys are completed during the "due diligence" phase of an acquisition program. Figure 2.2 is a more detailed pre-acquisition survey checklist used to collect information when making an audit of the physical facility. This form includes sections on repairs and maintenance contract review.

Figure 2.3 is a case study — a summary of results following a pre-acquisition survey and evaluation of a five-story office building.

Post-Occupancy Evaluation

Post-occupancy evaluations entail examining a building after it has been completed or occupied for a period of time. Information obtained from post-occupancy evaluations will document and verify that the performance of building systems (heating, ventilating, air-conditioning, electrical, etc.) meets the original design criteria or current space utilization requirements. It is important to ensure that these systems were properly commissioned after installation and that they continue to operate in conformance with the space requirements as they may change over time. Post-occupancy evaluations also help identify some of the strengths and weaknesses of buildings currently in use. They provide information for improving utilization of existing buildings and assist in determining requirements for future facilities.

Results of a post-occupancy building evaluation may include identification of indoor air pollution sources or energy-saving opportunities. These evaluations can document HVAC system changes or building modifications that may have had a negative impact on system performance. Results also document opportunities for reducing maintenance and repair costs and improving employee morale and productivity. Additionally, post-occupancy evaluations will help ensure that existing problems such as code violations,

Pre-Acquisition Evaluation Items

- ☐ Land cost per acre
- ☐ Size of the property
- ☐ Tax rates and structure
- ☐ Value of surrounding properties
- ☐ Soil conditions
- ☐ Availability/reliability of local utilities
- ☐ Available power capacity
- ☐ Types of power available
- ☐ Transportation
- ☐ Telephone
- ☐ Sewage
- ☐ Water supply
- ☐ Zoning
- ☐ Fire protection
- ☐ Sprinkler protection

- ☐ Labor supply
- ☐ Unions
- ☐ Community services
- ☐ Pollution regulations
- ☐ General business climate
- ☐ Diversity of business
- ☐ Easement restrictions
- ☐ Site topography
- ☐ Flooding
- ☐ Ceiling height (if existing facility)
- ☐ Number of truck docks
- ☐ Space to facilitate movement of vehicles and freight
- ☐ Parking space

Figure 2.1

Pre-Acquisition Survey Checklist

I. **Property Description**
A. General Description
B. Site
☐ 1. Flood insurance rating
☐ 2. Easements
   a. Ingress/egress
   b Storm sewer
   c Power & communications
☐ 3. Parking lot
   a. Drainage
   b. Lighting
   c. Code compliance
   d. Exclusivity
   e. Condition
   f. Ability to accommodate traffic flow
☐ 4 Water–availability, pressure & flow
☐ 5. Site lighting
☐ 6. Utility services
☐ 7. Landscaping
☐ 8. Sidewalks, driveways
☐ 9. Storm retention and detention
C. Building
☐ 1. Age
☐ 2. Exterior description & condition
☐ 3. Foundation
☐ 4. Construction type
☐ 5. Floor & roof load capacities
☐ 6. Caulking
☐ 7. Insulation
☐ 8. Window systems
☐ 9 Roof
   a. Construction
   b. Drainage
   c. Condition of roofing & flashing
☐ 10. Penthouse
☐ 11. Entrances
☐ 12. Handicapped access
☐ 13. Interior construction (in place)
   a Description
   b. Condition
☐ 14 Restrooms
☐ 15. Lobby
☐ 16. Lighting
☐ 17. Fire system
☐ 18. Elevators
☐ 19. HVAC equipment
   a. Air flow and balance
   b. Operation of fans, compressor, motors, bearings
   c. Sizing capacity

   d. Temperature & humidity control
   e. Zoning
   f. Ventilation
   g. Age & condition
☐ 20. Plumbing
   a. Water main
   b. Standpipe
   c. Wet columns
   d Hot water supply
   e. Drainage
   f. Sump pumps
   g. Water closets & lavatories
   h. ADA compliance requirements
☐ 21. Fire Protection & Life Safety Systems
   a. Alarms
   b. Extinguishers
   c. Fire pump & sprinklers
   d Code compliance
   e. Elevator
   f. Heat/smoke detectors
☐ 22. Electrical
   a. Transformers
   b. Amperage–capacity
   c. Breaker panels
   d. Bus ducts
   e. Lighting
   f. Metering
   g. Emergency feed–replacement options
   h. Age & condition
   i Security

II. **Condition & Estimate of Repair Costs**
A. Notations of Deferred Maintenance
III. **Maintenance**
A. Review of Maintenance & Service Contracts:
☐ 1. Window cleaning
☐ 2. Scavenger
☐ 3. Elevators
☐ 4 Landscaping
☐ 5 Janitorial
☐ 6. Security
☐ 7. Interior landscape
☐ 8. Fire alarm system
☐ 9. Snow removal
☐ 10. Mechanical–HVAC
☐ 11. Electrical
IV. **Code Compliance**
A. Fire and Life Safety
B. Building
C. Parking
D. ADA Compliance Requirements
E. Signage

Figure 2.2

bottlenecks (in physical layout or equipment use), and wasted space are not duplicated in future facilities.

Post-occupancy evaluations provide the basis for establishing performance benchmarks (more about this in Chapter 3) to compare one building's performance against other properties of similar quality and use, both within the organization and outside. These benchmarks help managers measure the progress of building improvements undertaken. The relatively small cost of a post-occupancy evaluation can be paid back many times over the life of a building through the resultant money-saving ideas, improved morale, and productivity of the workforce.

Functional Building Evaluations

Functional evaluations of facilities determine whether the facilities are fit for their intended use. They also indicate the feasibility (from a technical and economic standpoint) of making changes required to accommodate the organization's needs.

Many buildings on the market today that are aesthetically pleasing, are functionally obsolete or unable to accommodate basic needs. Ceiling

---

Pre-Acquisition Survey and Evaluation of Five-Story Office Building

Problem: To determine the general condition of the building and site, identify any remedial repairs or upgrades necessary to accommodate a planned program for use of the building, and prepare an estimate of fair costs for making needed improvements.

Observation: Building is typical of many speculative office buildings erected in the late 1970s and early 1980s. Some building systems were apparently designed with first-cost economies in mind (HVAC, electrical, and fire protection)

Recommendation: Upgrade and repair various building systems to maintain comfortable space conditions, reduce energy costs, meet new ventilation requirements, and extend the life of certain building systems. Remedial structural repairs are required to meet lateral wind load code requirements.

Result/Payoff: Pre-acquisition observations and recommendations, with estimates of probable construction costs for associated remedial repairs and improvements, assisted purchaser in arriving at an informed decision with respect to the purchase of the building, and helped the purchaser negotiate a substantial reduction from the seller's asking price.

Figure 2.3

heights may be limited. HVAC systems may be unable to meet standards for outside air ventilation and desired temperature conditions based upon modern office environments. Toilet facilities may be inadequate and unable to meet ADA requirements. Fire protection and other basic building systems may be insufficient.

Behavioral Building Evaluations Behavioral evaluations involve assessing the influence of buildings on the behavior of people. For example, the color of wall and floor finishes may affect office workers' emotions. Employees working in open office areas may find it difficult to concentrate with people walking around their work stations. Distracting peripheral conversation may be more interesting than the tasks at hand. Furthermore, it is difficult to talk about sensitive issues in open areas where private conversation may be heard or misinterpreted. Erratic noise and continuous high frequency sounds are stressful to some people and can cause headaches and ear problems. Both visual and noise distractions can decrease productivity and cause stress.

Technical Building Evaluations Technical evaluations involve assessing and describing the condition of physical building components and systems. Included are evaluations of structural, mechanical and electrical systems; building envelope; room finishes; life safety systems; and acoustics.

Structural Systems Soil borings and tests should be made to determine the bearing capacity and condition of the soils for foundation design. If you are evaluating an existing building, the thickness, strength and condition of the concrete floor slabs should be established to ensure proper conditions for proposed use or storage of heavy equipment, machinery, or materials.

In addition, the capacity and condition of the building's structural systems should be evaluated by a structural engineer. Costly modifications (such as floor, roof, framing, and foundations) may be required for support of machinery, equipment, and utilities.

Mechanical, Electrical and Other Building Systems The capacity of the mechanical and electrical systems should be investigated thoroughly to determine if they meet building codes and the requirements of its occupancy. Mechanical engineers and service technicians should make on-site inspections of the HVAC and plumbing systems.

A mechanical engineer can perform a thermal load analysis to determine if the building's heating and cooling systems are capable of meeting the new owner's requirements for space usage. The analysis determines if there is sufficient fan capacity to meet supply air quantity calculations. It also determines if existing heating and cooling equipment can maintain desired temperature and humidity conditions and zoning requirements. The mechanical engineer and technician also can alert you to possible code violations in the building's mechanical systems. Problems uncovered might include insufficient outside air to comply with ventilation requirements.

The existing roofing systems also should be inspected and tested. Some roofs may require major repairs, and others will need replacement. Consider the age and condition of the roof. The inspector should be looking for signs of decay, moisture penetration, and previous repairs to roofing and flashing. Bubbles, exposed roofing felts, and green plant growth are definite signs of roof deterioration, as are drainage problems, blisters, splits, moisture damage, deteriorated felts, and punctures.

An electrical systems analysis should be performed to determine specific requirements for electrical service, capacity, reliability, and needed modifications. A similar analysis will determine the adequacy of the plumbing and fire protection systems.

Also identify what is needed to comply with fire protection and life safety laws. Managers need to know the number and location of fire exits, panic hardware on doors, and facilities for disabled people. Investigate the need for smoke and heat detectors, annunciator panels, fire doors, fire alarms, automatic fire suppression systems and manual systems, and interior and exterior lights.

Architectural components should also be investigated. These include the building envelope, where cracks and openings in exterior walls may be sources for leaks and excessive heat gains and losses. Cracks may also be a sign of possible structural problems.

Space Requirements Space planning requirements can help determine if undeveloped land or existing buildings can accommodate your needs. Zoning regulations, exterior parking and traffic staging needs should also be considered. If your facility is serviced by "over the road" trucks, you need to provide ample space to accommodate them. Inadequate docking facilities can shut down a plant.

Environmental If your organization is considering purchase of an existing building or a site that may have been exposed to hazardous wastes, you must verify that no problems have been left behind by previous owners or users. Conduct a separate investigation of environmental concerns such as possible presence of asbestos-containing materials, underground storage tanks, and other potentially hazardous contaminants to the building environment. Facility managers cannot take chances in this area. If you are unable to prove that an existing industrial site is clean, it probably isn't.

Summary

Reliable pre-acquisition studies will help avoid the potential of investing in a project that is doomed to fail. They also will assist in determining the overall cost of a project.

Figures 2.4–2.8 are case history summaries of survey and analysis problems, observations, recommendations, and results/payoff for four different facilities.

Facilities evaluations are an essential element in the TPFM program. They help you establish your organization's program requirements, assess a proposed property, and evaluate a facility currently in use. The information these evaluations provide strengthens your ability to analyze and present factual, complete recommendations regarding property acquisition and improvements. The next step is to compare your facility's performance to that of other facilities within your own organization, and to outside organizations, through benchmarking.

Space Utilization Survey and Analysis of Existing Production and Warehouse Facility

The organization was considering leasing additional space in order to increase available square footage (beyond their current 75,000 S.F space) for expansion of their production operation A consultant was hired to evaluate existing space utilization to determine the additional square footage required. The consultant analyzed the existing layout and workflow and offered the following observations and recommendations.

1. Implement low-cost changes in material handling and storage.
2 Reconfigure work centers.
3 Combine certain operations into a central location.

With the above improvements, the proposed, expanded operation could be condensed into 38,000 S.F. of space. The excess 37,000 S.F. could be re-leased, and a relocation avoided

---

igure 2.4

Energy Management System Analysis

A consultant was retained to perform an energy analysis of 370 hotel guest rooms to determine their annual operating cost and the energy savings associated with a proposed installation of an energy management system.

The HVAC system for each room consists of a 3/4-ton fan coil unit having a chilled water cooling coil and a 2 KW electric strip heater. In addition, there is a 2 KW electric baseboard heater installed below the windows. The analysis is based on a computer model to match the projected monthly occupancies schedule

The analysis is based on maintaining the guest rooms at a temperature of 72 degrees year-round with the savings based on resetting the guest room temperature up to 80 degrees during the summer and down to 65 degrees during the winter, along with cycling of fan coil when the room is unoccupied.

The energy analysis concluded the following:

1. The annual operating cost for each guest room HVAC system is approximately $170, or $67,900 for the 370 rooms.
2. The total annual energy savings by installing the energy management system without fan cycling is approximately $6,666.
3. The total annual energy savings by installing the energy management system with fan cycling is approximately $9,537.
4. The energy management system savings did not conform to the corporate guidelines for a return on investment.
5. The energy analysis demonstrated that the greatest opportunity for saving energy and operating costs is in the public spaces of the hotel.

Figure 2.5

Rearrange Manufacturing Facility
Using Lean Production Concepts and Cell-Oriented
Continuous Improvement Teams

A leading manufacturer of water treatment products needed to improve manufacturing processes in order to increase productivity and optimize utilization of space. The organization also needed additional floor space in which to base operations purchased from another business. Budget constraints required that these goals be accomplished without a substantial investment in mechanization or automation.

A small group of manufacturing engineers, with the aid of supervisors, was asked to perform an evaluation of current operations They observed an opportunity to reconfigure manufacturing operations from traditional batch-and-queue techniques to lean manufacturing practices.*

Their recommendation

1 Group together machines and personnel that produce a family of related products — or parts with similar processing requirements — into work centers (cells) arranged in a compact, sequential manner.

When the cell system was implemented, inventory could be reduced along with associated carrying costs. Cycle times were also reduced, and material handling minimized. The reconfiguration frees up a substantial amount of floor space, enabling the company to successfully fit in the newly purchased operations. Reconfiguring production activities into work cells, focusing on the customer's needs, minimizes travel distances between functions and machines, enhances communication, improves teamwork, and eliminates the need to spend major capital funds on sophisticated automation equipment or new facilities.

*Examples of lean manufacturing techniques include: statistical process control and other quality management methods, inventory management, and workflow adjustments (such as changing from process layouts to focused work cells and machine centers). Other examples of lean manufacturing· high-involvement, team-based work methods, such as quality circles, self-directed work teams, and inexpensive changes to equipment and material handling procedures.

Figure 2.6

HVAC System Alterations

A consultant was retained to provide engineering services to perform energy/operational audits and to develop designs that would implement findings of the various studies. The facility is a commercial office complex consisting of a 10-story building (250,000 S.F.), 2–3-story office buildings (75,000–100,000 S.F.), and ground-level space (30,000 S.F of public and 55,000 S.F. of rental space), and an open parking garage.

The HVAC System recommendations were

1. Convert the existing constant volume air handling units with zoned reheat coils to a variable air volume distribution design without reheat.
2. As a result of the conversion of air handling units to a VAV design, the reheat system requirements are reduced. This allows the replacement of two 500 horsepower fire-tube boilers with two 95 horsepower water-tube boilers The design includes a separate domestic hot water system and a boiler hot water reset schedule.
3. The conversion of the air handling units to a VAV design also results in a reduction to the refrigeration requirements. This revision allows for modifications to the piping of the chillers and cooling towers to provide independent operation of each chiller at part load operation

With HVAC system alterations, the facility's gas and electric energy consumption can be reduced by over 50% annually, while maintaining tenant comfort.

Figure 2.7

Ventilation and Air-Conditioning
System Evaluation of 17-Story
Office, Retail, and Condominium Building

The facility was having difficulty maintaining desired temperature and air circulation conditions. A mechanical engineer analyzed the existing situation and made the following observations:

The floor air handling unit supply fans delivered approximately 22,000 CFM versus the 33,000 CFM intended by design. Inspection of the discharge duct at the fan outlet indicated a substantial "system effect" due to abrupt duct elbows, restricting performance of the supply fan. Static pressure measurements indicated that pressure drop across filters was excessive. Severe leaks were also observed in the medium pressure trunk duct.

The analysis resulted in the following recommendations:

1. Redesign the existing discharge duct for the maximum straight length to minimize the system effect.
2. Change from 85% efficient bag filters, with 211 throw-away type pre-filters, to 30-45% efficient filters normally used in office building HVAC systems.
3. Seal ductwork at leaking joints and fittings.

Taking the above actions will result in an approximate 33% increased air quantity delivered to spaces.

Figure 2.8

Chapter 3

Increasing Productivity Through Benchmarking

Just another buzzword? Everyone talks about *benchmarking*, but what does it really do for you? Benchmarking is a useful process for identifying strategic goals that may span a period from two to five years. Benchmarking helps to identify trends and the changes you need to make to be more competitive or to more effectively support the mission of the overall organization. Benchmarking identifies the gaps between your current practices (where you are now) and best practices (where you want to be). The secret to measuring facilities performance for benchmarking is developing quantitative indicators or standards for key building operations and real estate performance categories.

A benchmark is an *appropriate* point of reference. If the benchmark is not appropriate to your function or environment, it should not be used. A benchmark is an indicator of overall performance from a broad standpoint. By benchmarking, you may discover that your overall square footage of space usage is too high, or your downtime excessive. In the context of facility management, a benchmark refers to the *best* management techniques and technologies. In this text, the "best" refers to the state-of-the-art; that is, there is no better technology or system available. A simple example is changing the belts on HVAC equipment. Group belt replacement may be considered the best practice. If you are replacing four belts on an air handling system on an individual basis as they become overly worn or break, then you have a performance gap between your current method and what may be considered a best practice. By replacing all belts at the same time, you can avoid carrying a large inventory and eliminate emergency breakdown and unnecessary downtime.

So, the first principle is that a benchmark must be appropriate and represent the best method or technology for a given environment, process, or set of operating constraints. A facility manager should be able to "see" where his or her organization is compared to the benchmark. The facility manager must also be realistic—for example, not applying the same outside air exchange requirements as are suitable for a commercial office complex to an industrial process environment that requires removal of noxious, odorous, and toxic gaseous contaminants. While the specification for an office complex or warehouse space may be less costly from an operational standpoint, it is not appropriate.

Second, a facility manager must be able to assign a cost to "closing the gap" between the status of current operations and the desired benchmark. The tool to reach this goal is the **best affordable technology or methodology**. Everything the facility manager does for the facility must be presented in terms of dollar value, and adherence to life safety codes and government mandates. The manager must evaluate, "How much will it cost to close the gap between the current method and the benchmark, and how much is it worth?"

Third, benchmarking is a continuous process. It is important to keep in mind that the *process* may be just as important as the *product* of the benchmarking effort. Benchmarking may involve measuring your current business operations and comparing them with external values, quality, and reliability standards. It may include code, life safety, and insurance requirements. Benchmarks may be established for operating cost and environmental performance standards. Or, they may document the practices and costs of your toughest competitors or industry standards, and those of leaders in any industry performing similar functions. Benchmarking facilities performance involves looking at the best companies and finding out what they are doing better than you are.

While some benchmark information is available through facility management associations like IFMA (see address in Chapter 2 and publication information in the bibliography), and from technical associations like ASHRAE (American Society of Heating, Refrigeration, and Air Conditioning Engineers — Atlanta, GA), facility managers may, alternatively, choose to form a group and select benchmarking partners from the same or other industries. Together, they can determine appropriate benchmarks and best practices, and compare and challenge each other's performance. Benchmarking partners may be in similar, but noncompeting industries, with similar equipment and functions. Benchmarking partners may be in the same region, where utility and labor rates are the same. Facilities consultants can be helpful in putting together a good match. The data collected from the benchmarking process are invaluable in identifying company strengths and weaknesses and uncovering alternative practices which could lead to a competitive advantage.

The results of a benchmarking survey are useful for programming new facilities, evaluating existing facilities, comparing your facility performance with others, challenging your current ways of doing things, and setting performance targets for facility management functions. Benchmarking is a method facility managers can use to measure their performance and obtain data for targeting opportunities for improvement. Let your objectives drive the benchmarking study and determine what is to be benchmarked. Is your objective to:

- Minimize operating costs?
- Determine if you are spending money most efficiently?
- Reduce production downtime?
- Reduce cycle times?
- Save energy?
- Justify departmental budgets?
- Improve worker morale, health, safety, comfort, and/or productivity?
- Determine the need for capital expenditure?
- Decrease property insurance costs and Workers' Compensation premiums?
- Determine where to locate a new facility? (Regional operating cost factors may be a consideration, but could be secondary as a weighted average when compared with the organization's mission-critical objectives.)

Some benchmarking studies are less technical (operational) and more strategic (e.g., business trends such as telecommuting and outsourcing). Technical or operational benchmarking studies are more dollars-oriented, e.g., utility or janitorial costs per square foot, or square footage per occupant, or hours spent on preventive maintenance.

If managers wish to improve something, they must first be able to measure it or count it. There are tools and techniques that can be used to collect and organize data regarding facility performance. Statistical process control, occupant survey, ratio analysis and value engineering studies are methods that can be used to measure performance and establish benchmarks for improving the efficiency of facility operations.

Statistical Process Control

Once you have collected and organized your facility's current operating data, statistical principles can be applied to detect and record changes that affect quality, competitive edge, and productivity before the output becomes unacceptable. Statistical process control can be applied to any facility operation that has an identifiable, measurable output. The output might be air or water flow rates and temperature; humidity, dust or contaminant counts; equipment usage and repairs; production speeds; and maintenance. The manager must determine which performance criteria (output) to measure and then establish which variables to measure (as they affect performance). An example might be maintenance of room air between 70° and 80°F and the effect of discrete temperature and humidity variations within that range on production rates and product quality. The maximum change in temperature for certain R&D labs, telephone equipment rooms, data processing and reprographic spaces must not vary more than 5°F (plus or minus). The maximum change in humidity must not exceed 10% RH (plus or minus). Some production equipment, raw materials, finished goods, and building systems are more sensitive than people to environmental changes. These systems must operate within a controlled environment, requiring tighter controls for heating and cooling, air cleanliness, air flow and humidity control. The environment must be stable at all times.

The mean (average), median (middle value) and standard deviation (6 standard deviations account for 99.97% of all possible values for a given set of data) of the collected data is calculated and compared with performance levels or standards. Statistical process control can help uncover conditions or problem areas that affect processes and productivity.

Performance parameters should be selected that have the most impact on product quality, operating cost, and worker and machine productivity. Measuring the effect of temperature and humidity adherence on performance may lead to acceptance of less stringent temperature or humidity criteria or identify the importance of tighter HVAC (heating, ventilating, and air-conditioning) controls.

Statistical data may indicate facility operational areas requiring improvement, and areas where it is not cost-effective to make improvements. Cooling water and boiler chemistry; the number of unscheduled maintenance activities; the number of times needed parts are unavailable, late or unacceptable; and the number of defective parts are items that can be monitored to determine whether corrective action is necessary.

Figure 3.1 is a record of temperature and humidity variation over a three-week period for a press room space at a printing plant. The objective is to identify whether the performance target of 70° (plus or minus 2°F) and 50% relative humidity (plus or minus 5%) is being maintained in order to optimize productivity in the press room.

Statistical process control is based on **reliable processes**. Reliable processes are those that do not break down; therefore, effective statistical process control rests squarely on the shoulders of a **proactive** and **predictive maintenance** program. Processes must be accurate and repeatable. If a component in a process breaks down (for example: 99.95% acceptance level), all production must stop until the process element is repaired and an acceptable sample run of new products is made. Therefore, statistical process control as a technical philosophy must rely on proactive and predictive maintenance to prevent breakdowns. Preventive maintenance is not broad enough to fulfill the needs of statistical process control.

Week of 6/14/93

TEMP 71°F +/- 2°F

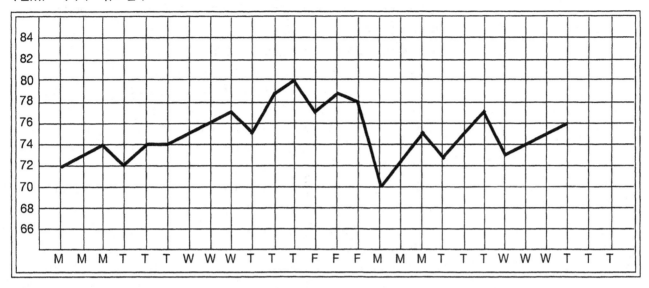

R.H. 50% +/- 5%

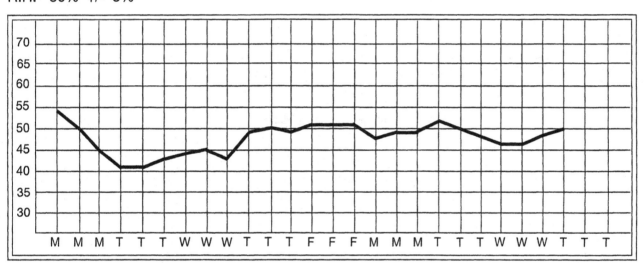

Figure 3.1

Occupant Survey Method

Building performance can also be measured through occupancy surveys. The main source of information about poor building performance is usually from occupant complaints. Individual departments or groups within buildings have their own performance needs and should be surveyed separately.

Ratings for different categories serve as useful feedback from occupants regarding their building performance priorities and needs. Examples include the need for more space; privacy and storage; improved lighting or ventilation; or support facilities for voice, data, and video communications. Also, survey results are useful for setting operating and performance standards which should be monitored continuously.

Figure 3.2 is an example of a Building Occupant Survey format.

You might classify the data collected by department or air-conditioning zone in order to identify real needs in certain physical areas. One of the drawbacks of this type of survey is that the information is subjective, and general rather than specific. However, this general survey may signal the need for a more costly and time-consuming engineering analysis and collection of quantified support data.

Ratio Analysis Method

Another useful method for measuring overall building performance is the use of "ratio analysis." Ratios can be developed for significant building operations and performance categories.

Ratio analysis can be used to compare a facility's current performance with its own previous performance and with the results of other firms. The indices may identify undesirable trends or areas to be corrected.

To use the ratio analysis method, standards should be developed for the most important facility management performance areas. Measurements should be regularly taken and compared with available standards. If none are available, you may need to develop your own.

Use of ratio analysis raises questions that lead to cost-saving answers For example, how can we reduce the power required to make a product? Do certain operations take up more space than necessary? Will additional space become available? Ratios should be calculated regularly to provide the feedback necessary to measure performance over time. Ratios also can be plotted on graphs to illustrate performance trends.

Value Engineering Studies

The term "evaluation" means to determine or judge the value, or worth, of something. Value engineering can be employed as a tool for facility evaluation, cost reduction and for the development of building performance standards. Value engineering studies on building systems and facility services are necessary to ensure that they are designed, constructed and operated in the most economical way, considering life cycle costs.

44

Building Performance Review
Occupant Survey

		Poor	Fair	Good	Excellent
1 Indoor Air Quality					
A	Air Freshness	1	2	3	4
B	Other, specify	1	2	3	4
2. Thermal					
A	Air Temperature				
	Winter	1	2	3	4
	Spring	1	2	3	4
	Summer	1	2	3	4
	Fall	1	2	3	4
B	Humidity				
	Winter	1	2	3	4
	Spring	1	2	3	4
	Summer	1	2	3	4
	Fall	1	2	3	4
C	Air Movement	1	2	3	4
D	Other, specify	1	2	3	4
3 Illumination					
A	Electric lighting	1	2	3	4
B	Daylight	1	2	3	4
C	Glare	1	2	3	4
D	Brightness	1	2	3	4
E	Color	1	2	3	4
F	Other, specify	1	2	3	4
4. Acoustics					
A	Noise levels	1	2	3	4
B	Voice privacy	1	2	3	4
C	Specific sounds of voices	1	2	3	4
	and equipment	1	2	3	4
D	Noise distractors	1	2	3	4
E	Noise from air systems	1	2	3	4
F	Exterior noises	1	2	3	4
G	Other, specify	1	2	3	4
5. Spatial					
A	Amount of space	1	2	3	4
B	Relationship of spaces/layout	1	2	3	4
C	Sanitary services	1	2	3	4
D	Privacy	1	2	3	4
E	Handicap accessibility	1	2	3	4
F	Aesthetic quality of exterior	1	2	3	4
G	Aesthetic quality of interior	1	2	3	4
H	Image	1	2	3	4
I	Adaptability to changing uses	1	2	3	4
J	Furniture comfort	1	2	3	4
K	Other, specify	1	2	3	4
6 Architectural and Interior Design					
A	Ceiling height	1	2	3	4
B	Quality of materials	1	2	3	4
	Walls	1	2	3	4
	Floors	1	2	3	4
	Ceiling	1	2	3	4
C	Color, texture, ornamentation	1	2	3	4
D	Security	1	2	3	4
E	Other, specify	1	2	3	4
7 Maintenance					
A	Mechanical systems	1	2	3	4
B	General housekeeping	1	2	3	4
C	Lighting	1	2	3	4
D	Parking lot	1	2	3	4
E	Roof	1	2	3	4
F	Building structure	1	2	3	4
G	Other, specify	1	2	3	4

Circle the appropriate rating (Poor, Fair, Good, or Excellent) to indicate your opinion of these facility conditions

Figure 3.2

The approach is to examine the building systems and services that require the greatest expenditures, to challenge the costs and functions, and compare them with alternatives on a cost and benefit basis. Value analysis is a systematic team approach to problem solving based on the analysis of functions, to identify and eliminate unnecessary costs without sacrificing quality or performance. Value engineering is addressed in more detail in Chapter 4.

The Benchmarking Process

All too often, management fails to systematically evaluate their facility resources and services on a periodic basis for: operating economics and efficiencies, building defects, productivity, waste, worker satisfaction and comfort, and code compliance. Benchmarking is a methodology that can help to organize the process and provide meaningful data.

Following are 10 key categories of facility management performance in terms of their impact on profitability and exposure to financial risk. The next part of this chapter explores specific benchmarking techniques for each of the above categories.

- Space utilization
- Energy costs
- Waste minimization
- Maintenance costs
- Indoor air quality
- Thermal quality
- Illumination quality
- Acoustical quality
- Ergonomics and safety
- Project management

Benchmarking: 6 Basic Steps

- Identify and scope benchmarking projects
- Identify standards and practices as a measure for facilities performance
- Collect and organize data
- Determine current performance variance
- Determine the cost to reach a higher standard, and how much it is worth
- Develop a framework for project implementation

Space Utilization

Are you paying more for space than you need to? Do you occupy more space than is necessary? Is the arrangement of equipment and departments efficient with respect to workflow (movement of materials and people)? Are you using space efficiently? Or just filling it up?

Good space utilization means direct and continuous flow of materials that minimizes unnecessary travel distances, backtracking, and other wasted motions; sufficient space around operations for work-in-process, maintenance and safety; adequate space for storage and handling of materials; unobstructed aisles and fire exits; and physical separation of spaces requiring different environmental conditions and building use classifications. Good space utilization maximizes use of available horizontal and vertical space. Utilizing a vertical storage system with racks reduces floor area requirements. Keep materials close to where they are needed by strategically locating staging areas.

Every facility should be analyzed on a periodic basis from a space utilization standpoint. This is one way to ensure that the necessary levels of service are provided in a manner that is effective and compatible with business objectives. Figure 3.3 shows ratios that may be helpful in determining whether space is being used in the most efficient way. While there clearly is a trend toward reduced space per employee, innovations such as systems furniture, open office plans, high-density file systems, digitizing of documents (rather than storing hard copy), and telecommuting, have reduced the need for space. Improvements in space utilization may result in an increase in the number of building users or occupants per square foot. As a result, HVAC systems should be re-evaluated to ensure that they can maintain temperature control and code ventilation requirements. Figure 3.4 is a sample Space Utilization Analysis, showing the current space allocations versus the space actually needed for a production and warehouse facility.

Department managers should be asked to complete questionnaires, charts, and forms to indicate critical requirements for space layout, utility, temperature, humidity, air cleanliness, ventilation/exhaust, air velocity, air pressurization, plumbing and sewage, power and lighting, voice and data, and other support systems. Secure a complete list of all equipment used in various operations. Include condition, age, and plans for replacement, as well as production and cost statistics.

Facility requirements are based on sales projections, environmental laws and trends, production needs, productivity rates, type of equipment, and usage. Anticipated growth rates and objectives over a 3–10 year period should be covered in the analysis. Strategies should be based on different business scenarios which could develop: optimistic, pessimistic, and most likely.

The Flow Chart To achieve significant, sustainable improvements you need to understand the way work is presently being accomplished and recognize opportunities for improvement. A team looking for ways to improve a process often creates a flow chart of that process as a first step.

Start by preparing an "as is" process flow chart. Doing this serves as a tool to document and describe the current process. The "as is" flow chart helps work teams identify bottlenecks, unnecessary or wasted

Space Utilization Ratios

$$\frac{\text{Production area}}{\text{Total facility area}}$$

This indicates whether valuable space is being used to accommodate low revenue producing operations. It can be applied to each department that occupies space within a facility Perhaps there might be a better use for the space available

$$\frac{\text{Area required for a particular operation}}{\text{Total production area}}$$

(see explanation above)

$$\frac{\text{Sales}}{\text{Square foot}}$$

A benchmarking comparison of annual revenue generated by different departments or divisions within a single organization (or by other similar companies) per square foot

$$\frac{\text{Mortgage/rental cost}}{\text{Square foot}}$$

A comparison of your organization's mortgage or rental costs with those of other divisions or facilities.

$$\frac{\text{Property Tax}}{\text{Square foot}}$$

Based on benchmarking data, a comparison of the annual property tax paid by printing companies ranged from $.31 per square foot to $1 74 per square foot.

$$\frac{\text{Net usable space}}{\text{Total space available}}$$

The usable area (as opposed to gross or total area) of a facility measures the actual occupiable area It excludes circulation halls, lobbies, toilet rooms and custodial areas.

$$\frac{\text{Employees}}{\text{Square foot}}$$

For example, the gross (total) amount of space per employee in an industrial setting, in this case the printing industry, may range from 1,175 square feet to over 3,166 square feet For offices, average space per employee continues to decrease, currently from 73–280 S.F (from *Benchmarks III*, published by IFMA in 1997) Local building codes may mandate minimum space requirements.

Figure 3.3

Sample Space Utilization Analysis

Ident.	Subject	Current Sq. Ft.	Percent Utilized	Required Space (Sq. Ft.)	Notes
A	S/R Dock	2,790	66%	1,800	Combine A-S
B	S/R Stage	1,815	N/C	1,815	
C	Bulk Storage	1,270	N/C	-	Put in Racks
D	Maint. & Mtrl.	2,890	50%	1,450	See Notes
E	General Office	5,076	N/C	-	See Notes
F	Warehouse	1,422	33%	500	Combine F-I
G	Packing	2,618	N/C	2,618	
H	Kit Assembly	1,029	N/C	1,029	
I	S/R Clerk	280	N/C	280	Combine F-I
J	Bulk Pallet Rack	9,316	80%	10,000	See Notes
K	Pick Shelves	3,900	75%	2,650	Aisles & Shelf
L	Print Shop	3,456	N/C	3,456	Store Paper
M	Incentive Store	962	50%	500	
N	Obsolete Pick	2,772	Delete	-	
O	Bulk Storage	3,672	66%	2,500	Use Racks
P	Archives	1,200	75%	1,050	Aisles Only
Q	Bulk Storage	9,824	25%	2,340	Use Racks
R	Mail Room	5,280	N/C	5,280	Integrate Strge.
S	Trash Dock	2,340	35%	780	Combine A-S
T	Dock Storage	930	Delete	-	
	USABLE TOTAL SQ. FT.	62,842		38,048	
	AISLES*	12,568 Aisles	15%	5,707	
	Total Square Footage currently in use	**75,410**		**43,755**	**Required Space**

*Total square footage of the aisles as a percentage of the usable space in the building @ 20% is not appropriate for low velocity product flow.

Delete = Space no longer needed for this function.

N/C = No change required; current use of space is optimal.

This analysis was for an existing light assembly production.

motions, redundant or other nonvalue-added steps related to the current method of getting the job done. After the team has analyzed the current process and determined the changes needed to achieve the desired improvements, a flow chart of what the process *should* be must be developed.

Process Flow Chart There are different kinds of flow charts. A basic process flow chart represents what happens at each step in a process. Figure 3.5 identifies the tasks, as well as the sequence and timing of each task, to process a specified job at a printing company. Consider developing detailed charts for operations within individual departments to obtain a better understanding of the flow of work within them. Details such as cycle times, number of movements between operations, or value added content can be added to the flow chart depending on the team's objectives. It may be possible to save time, money, and space by reducing the number of queues; eliminating or combining operations; reducing the number of times material is handled; and eliminating duplication of effort. Consider specialized equipment to improve productivity and space utilization, such as automated material handling systems, racking and shelving systems, high-density file systems, and electronic document storage.

Deployment Flow Chart A deployment flow chart basically identifies who is supposed to do what, and when, and how each department's responsibilities relate to the overall process. Figure 3.6 is an example.

Workflow Diagram Workflow diagrams can be used to identify the movement of people walking around operations, the flow of documents through an office, or the routing of materials through a production operation. Preparing a workflow diagram often helps to identify space layout changes that will eliminate wasted motion or inefficient use of space. Figure 3.7a is an example.

The workflow diagram is created by drawing a plan identifying operations at each step of the process under study. A line is drawn or the plan identifying each movement, and the relationship and sequence of operations. Additional lines can be drawn between operations to identify the number of trips or intensity of workflow between those operations. (See Figure 3.7b, an Adjacency Relationships Chart.) The prime purpose of the workflow diagram from a space utilization and layout standpoint is to show the importance of activity relationships, and to find ways to minimize delays, backtracking, and travel time.

Bubble Diagram The hand-drawn bubble diagram (Figure 3.7c) is a preliminary planning tool to define possible flow, space, and activity relationships for space planning.

Energy Costs

Are you paying more for electricity and fuel than is necessary? Are your facilities energy-efficient? As energy is used more efficiently, product cost can be reduced and profits improved. An energy audit is

Figure 3.5

Task ID	Task Description	Cycle Time Hours	Departmental Resource (Equipment)	Total Float
10	Order and Receive Stock	24		0
20	Condition Stock	24		208
30	Make Plates	32	Plate-maker	0
40	Queue	8		0
50	Move Stock to Press	2		0
60	Press Make Ready	8	Press #4	0
70	Print	8	Press #4	0
80	Queue to Bindery	8		0
90	Move to Bindery	2		0
100	Make Ready: Stitching, Shrinkwrap	6	Shrink-wrap	0
110	Stitch & Shrinkwrap	32	Stitcher #1	0
120	Move to Loading Dock	8		0
130	Ship Product	16		0

Legend:
- Early Bar
- Float Bar
- Progress Bar
- Critical Activity

FLOW

SAMPLE PRINTING CO.
JOB WORKFLOW - PRINTING OPERATION
SCHEDULE BASED ON ONE 8 HOUR SHIFT/DAY

Sheet 1 of 1

Project Start	27AUG98 00 00
Project Finish	23SEP98 08 59
Data Date	27AUG98 07 00
Plot Date	22OCT98 15 00

Deployment Flow Chart—Screen Printing Operation

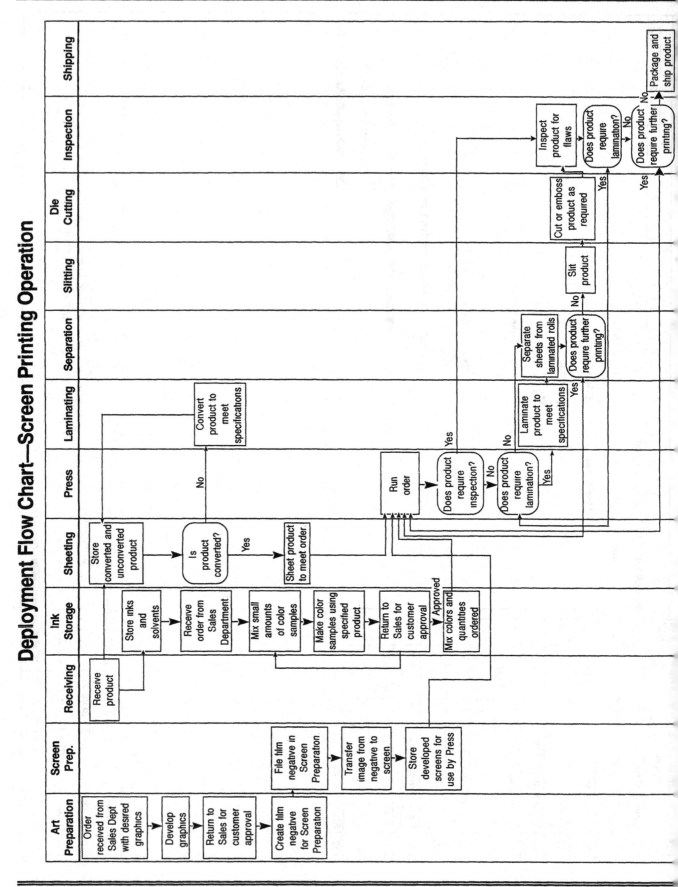

Figure 3.6

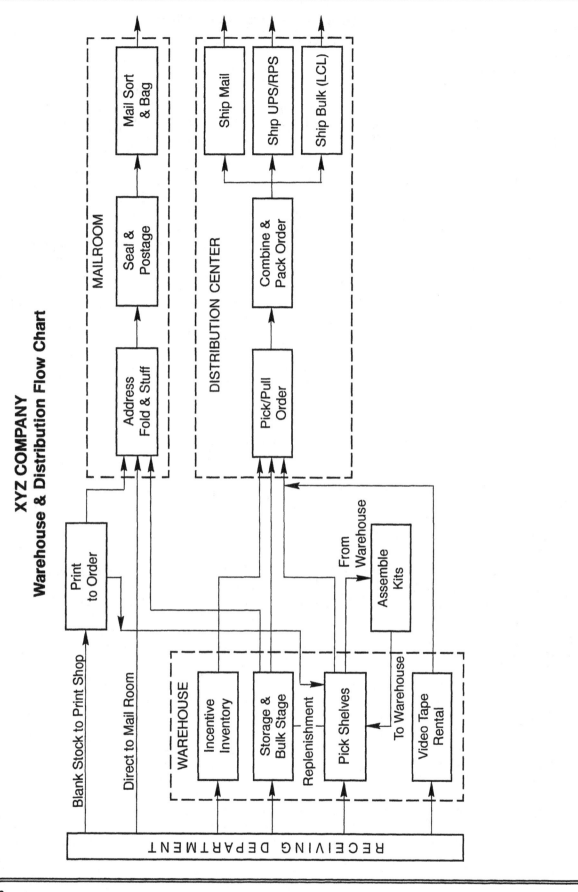

XYZ COMPANY
Warehouse & Distribution Flow Chart

Figure 3.7a

Adjacency Relationships Chart

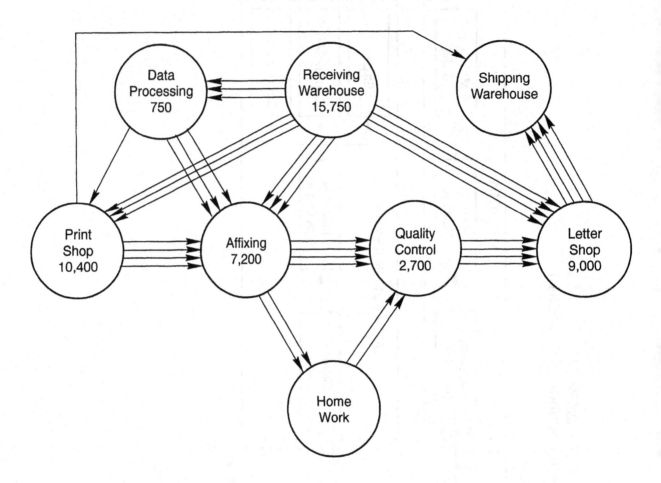

The number of lines visually represents the intensity of workflow between various production operations in this printing plant. It is a useful tool for analyzing the importance of adjacency relationships between various areas when designing or re-designing a layout. It can be applied to describe the importance of closeness between persons, activities or departments.

Figure 3.7b

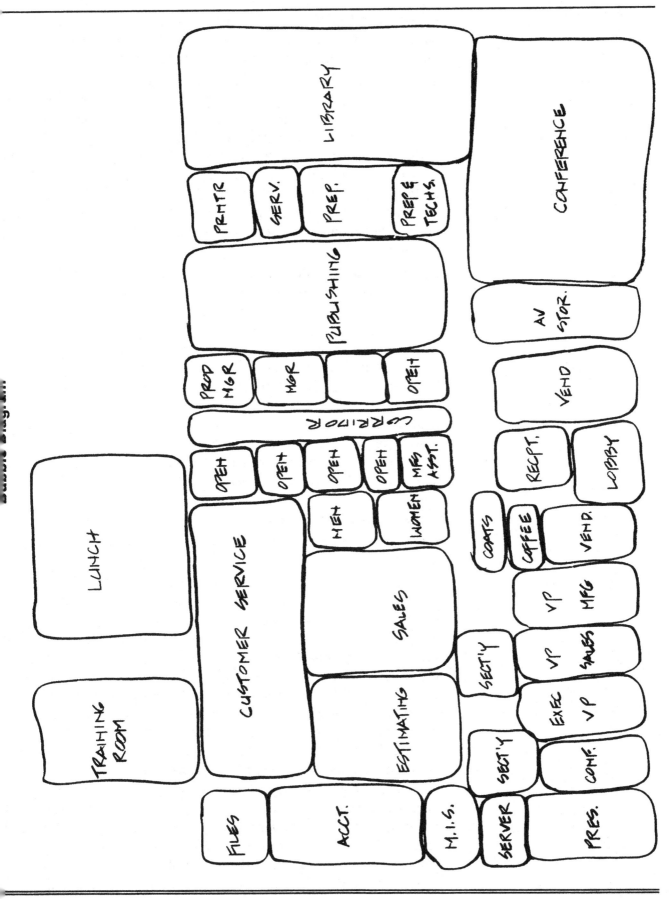

Figure 3.7c

necessary to identify energy consumers within your building, understand specific energy usage patterns within the building, and identify opportunities for reducing energy consumption.

Measurable goals should be set for reducing facility energy consumption. An orderly accounting of energy used in a building should be compared against some standard of performance or budget for example, kilowatts of energy consumed, or cost per given time period. Opportunities for energy conservation can typically be found in HVAC equipment, air distribution systems, temperature control systems, electrical distribution and lighting systems, and production equipment. Heat losses and gains through floors, walls, ceilings, roofs, doors, and windows, and occupancy levels and activities also affect energy use.

Significant energy savings often may be realized with relatively minor modifications or investment. Changes in operation or maintenance procedures may be all that is necessary to achieve significant energy savings. Recalibration of controls, regular filter changes, and coil cleaning are examples of simple procedures that save energy. The energy conservation measure may also be as simple as shutting off lights and setting back temperatures during unoccupied periods. Reduced lighting levels, and using energy-saving motors and luminaries or task lighting can reduce energy consumption as well. If a pump or fan is running improperly, adjustments will save money. Sometimes, larger energy-consuming air handling equipment and systems are operated during evenings or weekend unoccupied periods when a smaller, more efficient system could be used instead to treat a specific area, such as a computer or break room.

Figure 3.8 is an outline of procedures for conducting a facilities energy audit.

Some heating and cooling systems may have been oversized back when energy costs were relatively inexpensive. In other situations, in order to reduce front-end engineering costs and minimize the risk of under-sizing, engineers may have been inclined to oversize systems to avoid the analysis necessary for optimum system sizing and efficient operation. Boilers, pumps, furnaces, fans and motors may all have been oversized. Heating and cooling loads should be calculated to determine how much heating and cooling must actually be produced to meet the space loads and to condition ventilation air. Heat generated from lights, people, and equipment is a primary source of indoor cooling loads. Duct distribution systems should be examined for proper sizing, layout, and air balancing. Duct leakage should be minimal. Leaking water, steam, or inert gas may seem quite small as it escapes into the air, but over time it can represent a sizable amount of energy. Figure 3.9 shows energy performance ratios useful in conducting an audit and analyzing the results.

Today there are a variety of sophisticated energy cost-controlling devices. These include optimum start/stop of equipment, variable

Energy-Operational Audit (Typical Procedures)

I. Initial Review

    A. Building record drawings
- 1 Mechanical
- 2 Controls
- 3 Electrical
- 4 Architectural

    B. Utility records and operating logs
- 1 Gas
- 2 Electric
- 3 Oil
- 4 Operation logs

Note These records should be reviewed for at least the two previous years

    C. Utilization
- 1 Present usage
- 2 Operating schedule
- 3 Environmental requirements for space conditioning

Note For some facilities it may be necessary to electronically monitor the electrical power usage and demand characteristics

    D. Maintenance and repair records for all major pieces of equipment

II. System Review–Office Phase

    A. Review records and documents
- 1 Building record drawings
- 2 Utility records
- 3 Maintenance records

    B. Determine the type and kind of mechanical equipment and controls intended to serve specific areas, as well as the characteristics and limitations of the equipment and controls

    C. Compare original equipment design intent to equipment capability and operating requirements

    D. Analyze HVAC systems with regard to comfort or process conditioning criteria and load handling capability
- 1 Part load performance
- 2 Full load performance

    E. Evaluate controls, operating sequences and control parameters

    F. Interface with utility companies
- 1 Discuss systems revisions to achieve more favorable rate structures
- 2 Review economic feasibility of converting from one energy source to another

III. System Review–Field Phase

    A. Survey property for comformance to plans
- 1 Mechanical and equipment schedule
- 2 Electrical
- 3 Architectural

    B. Interview management and operating personnel to determine inadequacies of present systems and controls

    C. Perform an in-depth site survey of each operating system

    D. Conduct operational tests to validate the performance of controls, equipment and systems as required

    E. Perform selected field measurements with engineering test equipment

IV. Analysis and Report Phase

    A. Perform preliminary load calculations to compare system capabilities with present demand

    B. Interview management and operating personnel to determine inadequacies of present systems and controls
- 1 Automated controls
- 2 Mechanical equipment status
- 3 Energy utilization
- 4 Useful economic life vs present age and condition

    C. Identify alternate control sequences for mechanical systems
- 1 Intermittent operation based on time clock and/or temperature parameters

    D. Mechanical services
- 1 Preventive maintenance
- 2 Revision or remedial work on systems and equipment

Figure 3.8

frequency drives which reduce fan speeds when demand decreases, outdoor economizers (for free cooling), electrical demand charge limiting, computerized energy management systems, duty cycling, night setbacks, chiller optimization, and process heat recovery systems.

You will want to set goals for reduction of energy usage based on the audit and selecting the energy conservation measures with the highest payback. Once an energy conservation program is established, building systems will need a routine examination to monitor energy usage and ensure efficiency.

The Effects of Utility Deregulation on Energy Costs

The deregulation of utilities will be an increasingly important development for facility managers as competition is introduced into these businesses. In the old structure, electric utility companies owned their own power plants, transmission systems, and distribution networks, and exclusively covered a certain geographic region. In the new system, the three components are separated, transmission and distribution continue to be regulated, and generation (deregulated) is

---

Energy Performance Ratios

$$\frac{\text{Kilowatt-Hours}}{\text{Square Foot}}$$

A month-by-month record of a building's electric energy use pinpoints the times when energy use is largest and therefore energy saving opportunities are greatest

$$\frac{\text{Square Foot}}{\text{Cooling (Tons of Air Conditioning)}}$$

Data needed to determine cooling capacity requirements include production equipment heat rejection rates, occupant levels and activities, quantity of outside air for exhaust make-up and ventilation, humidity control, indoor temperature requirements and heat gain through the building envelope. Cooling expressed in tons of air conditioning represents a substantial percentage of overall energy cost. According to IFMA's *Benchmarks II Report*, for commercial offices, the average cooling requirement is 47 S F. per cooling ton. The average monthly utility consumption is 2.2 kilowatt-hours per S F (Actual figures vary significantly by geographical region, type of building construction, and occupant density)

$$\frac{\text{Electric Cost}}{\text{Square Foot}}$$

For example purposes, energy costs may range from $3 to over $6 per square foot. Energy cost may represent from 1% to over 4% of sales revenues for a printer

Figure 3.9

available from more than one producer or broker. The other option is continuing to purchase power from the same local utility, with the three components bundled together.

The natural gas industry has already been similarly deregulated in some areas, where customers purchase from a selected provider. The gas is transported from a distribution point by the local utility.

It is expected that deregulation will bring about significant savings to energy purchasers due to better efficiencies and an ability to select not only providers, but specific services. Legislation is in development for the setting of deregulation guidelines.

Waste Minimization

Waste may be generated through production processes, maintenance operations, material storage and handling procedures, finished product, and work in process. Management should set policies and strategies for minimizing both hazardous and nonhazardous waste. There are legal and economic incentives to reduce or eliminate waste. The United States EPA recommends that a Waste Minimization Opportunity Assessment, sometimes called a *waste minimization audit*, be conducted periodically by facility owners and managers. The assessment provides valuable direction on visible and less detectable violations of local, state, and federal laws. The assessment consists of a careful review of a plant's operations and waste streams, waste measurements, disposal methods, segregation processes, disposal costs, and identification of alternative methods for disposal.

One of the most effective methods for reducing waste is reducing energy input into the process. This means doing a mass balance equation for the process — an energy analysis.

    Input includes: material, labor, and energy.
    Output includes: finished product and waste.
    Input must equal output; no other choice is available.

Since material requirements may be fixed for a given output, it may not be possible to reduce the material input. Since labor may be considered as an extension of equipment (operators feed stock or manipulate levers, and so forth), it may not be possible to reduce labor without reducing product output. That leaves energy as the only specific input that can be reduced without reducing production. It takes energy to produce waste. Therefore, the mass balance equation — input equals output.

Calculating the amount of energy required to produce the output, and subtracting that quantity from the amount of energy used yields the amount of energy available for savings. This leads to reduced waste, since product output is not reduced. Again, no other choice is available.

The facility manager begins this analysis by identifying how much material, labor, and energy go into the process and comparing these amounts to how much is actually needed (see Figure 3.10).

The quantities needed are measured by conventional technology. Figure 3.11 shows waste minimization ratios that may be helpful in conducting this evaluation. It is useful to prepare flow charts identifying the source, type, and rate of waste-generating processes and accounting for losses and emissions. Once the waste streams are established, options to minimize waste (such as source reduction or recycling) can be developed and evaluated. The feasibility of the options should be evaluated from a technical, economic, and legal standpoint. Standard equipment, such as predictive maintenance technologies, are suitable for identifying specific modes of process energy loss in the form of excessive vibration, temperature, noise, and misalignment.

Treating and recycling the waste back through the manufacturing process is a method of recovering the energy content of waste. It may be possible to burn the waste and use the recovered heat in the process [for example, from the thermal incineration of volatile organic compounds (VOC's) from production operations]. If there is sufficient waste heat available, it is possible to route the waste heat to a heat recovery chiller to provide chilled water for process requirements and for plant air-conditioning. If the treated exhaust air is clean and within safety limits specified by OSHA and other applicable regulatory bodies, it may be feasible to recirculate the exhausted air, thus saving the energy it would take to heat or cool additional make-up outside air. Safety features can be built into the system which automatically revert to full exhaust in the event of a system failure.

Material handling is another source of excess energy input. Moving material from place to place is wasted work and does not add value to any product; it only increases the probability of damage to products and consumes additional energy.

Maintenance Costs

Here are some questions every facility manager should be able to answer. Are you paying more for maintenance than you need to? Are you investing enough money in maintenance? What is the condition of your equipment? What percentage of your maintenance is planned? Are you doing enough of the proper type of maintenance? Do records exist that document equipment maintenance history?

Your business may be able to afford to purchase an asset only once, so it is wise to care for it properly. A facility manager needs to set up an effective maintenance organization and establish appropriate maintenance procedures. These efforts should consist of regular inspections and effective routine practices that will:

- Eliminate causes of failure
- Prolong equipment life
- Prolong the life of the facility
- Reduce the probability of costly major breakdowns, emergency repairs, and lost production

General Flowsheet for Lithographic Printing
Artwork, Copy, or Other Image

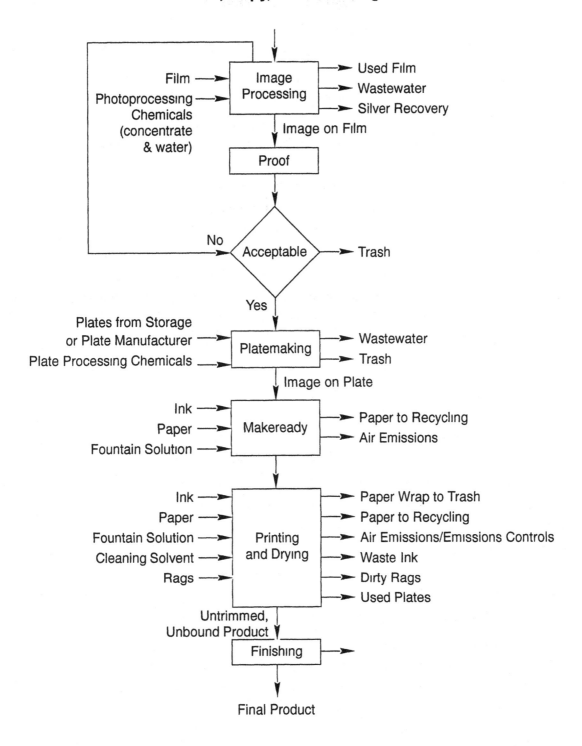

Prepared by the EPA in their *Guide to Pollution Prevention–
The Commercial Printing Industry*, August 1990)

Figure 3.10

- Reduce operating costs
- Improve company image
- Decrease downtime
- Ensure safety
- Ensure smooth operation of equipment at peak efficiency
- Save energy
- Maintain warranties

Maintenance represents a growing portion of many facility budgets and can be broken down into four broad categories:

- Proactive Maintenance
- Predictive Maintenance
- Preventive Maintenance
- Breakdown or Emergency Maintenance

1. Proactive Maintenance includes those activities aimed towards eliminating the root causes of equipment failure, thereby extending equipment life indefinitely. Proactive maintenance focuses on two primary root causes of equipment failure: excessive heat and dirt. These elements operate together to cause failure of equipment. Dirt

---

Waste Minimization Ratios

$$\frac{\text{Annual Cost of Nonhazardous Trash Removal}}{\text{S.F.}}$$

$$\frac{\text{Annual Cost of Hazardous Trash Removal}}{\text{S.F.}}$$

For example, in the printing industry:

$$\frac{\text{Incoming Pounds of Through-Put Paper}}{\text{Outgoing Pounds of Waste Paper}}$$

These ratios are useful in evaluating opportunities for recycling to reduce costs.

Figure 3.11

acts as an abrasive that causes parts to wear, and as an insulator that does not permit heat dissipation — a lethal combination for equipment.

Proactive maintenance seeks to establish maximum levels of acceptable equipment temperature and cleanliness. Then activities are planned and scheduled to ensure that the maximum acceptable levels of temperature, cleanliness, and alignment are not exceeded.

Most often, equipment fluids (lubricants, hydraulic fluids and heat transfer fluids) are the focus of proactive maintenance. A level of cleanliness is established for those fluids, and dirt is filtered out of the operating environment. The environment is tested to ensure that the fluids are clean. Fluids are replaced if they are too dirty.

These maintenance practices have been proven to be the most cost-effective over time. See Figure 3.12a for a comparison of maintenance strategies. Studies on bearings such as the SKF Infinite Life Study, have proven that these components can, in fact, have "infinite life" if lubricants are kept clean. Under proactive maintenance the maintenance staff *verifies* that equipment is clean, properly aligned, runs cool, and has clean fluids. Figure 3.12b illustrates the four different maintenance strategies, comparing their cost per horsepower per year, and using the human body as an analogy.

2. Predictive Maintenance includes those activities and technologies that detect pending equipment failure. Predictive maintenance is based on equipment usage and associated component wear and remaining part life.

Typically, vibration analysis, oil analysis, thermography (infrared scanning), ultrasonic testing, and electrical current and waveform analysis are the primary technologies. Vibration analysis reveals how much wear a part has experienced and when it is likely to fail. Oil analysis shows which parts are wearing and how much. Thermography measures how hot a part or system is.

Ultrasonic testing is used to detect excessive noise levels in equipment such as failing steam traps, bad bearings, and corona discharge (arcing in electrical components). Electrical analysis detects faults and pending faults in electrical equipment such as switchgear, motors, and control systems. These technologies allow facility managers to compare existing performance with appropriate levels of operational parameters. For example, a facility manager using vibration analysis can tell how much a bearing is vibrating and compare that to how much it should be vibrating.

When a component is found that is likely to fail, maintenance work can be planned and scheduled to replace that component. This philosophy precludes unscheduled equipment failure and the associated unexpected downtime.

3. Preventive Maintenance includes activities that are performed routinely at planned time intervals on specific systems, equipment, or

Machine Maintenance Strategies

Proactive Maintenance Conditions are maintained that avoid the onset of machine wear and component degradation. Conditions are monitored and remedial actions are taken to stabilize healthy operating conditions Maintenance activities are proactive, i.e., ahead of the first initiation of failure, not in response to it.

Predictive Maintenance The progression of failure is monitored using nondestructive instrumentation. Machine repair is scheduled prior to catastrophic breakdown.

Preventive Maintenance Maintenance is scheduled according to historic trends, experience, or reliability data. Typically, operating intervals such as hours, miles, or cycles are used as a basis for maintenance, not machine condition. Considerable guesswork is involved

Breakdown Maintenance Maintenance is scheduled in response to operational failure.

(Adapted from *DIAGNETICS, INC.,* by James Fitch, Tulsa, OK)

Figure 3.12a

The Human Body Parallel to Machine Maintenance

Maintenance Strategy	Technique Needed	Cost per HP per Year	Human Body Parallel
Proactive Maintenance	Monitoring and correction of failure root causes, e.g., contamination	$ 10	Cholesterol and blood pressure monitoring with diet control
Predictive Maintenance	Monitoring of vibration, heat, alignment, wear, debris	$8	Detection of heart disease using EKG or ultrasonics
Preventive Maintenance	Periodic component replacement	$13	Bypass or transplant surgery
Breakdown Maintenance	Large maintenance budget	$18	Heart attack or stroke/hospitalization

Figure 3.12b

components in order to avoid total failure. Examples include changing filters, oils, and drive belts. This philosophy does not account for levels of equipment usage or wear. In fact, components are often changed or replaced when they still have useful life.

4. Breakdown or Emergency Maintenance, also known as *remedial maintenance*, is an activity performed on a nonroutine basis and is reconstructive in nature. The equipment runs until it breaks. When there is a breakdown, maintenance crews are sent to "troubleshoot" the problem and "fix it." Repairs under this philosophy can, and often do, include everything from applying *"Band-Aids"* to complete equipment rebuilds. Estimates of costs and downtime are usually suspect since spare parts and skilled maintenance technicians may or may not be available.

Equipment that is allowed to run until it breaks is always subject to *extraneous dynamic loading*. Essentially, when equipment experiences extraneous dynamic loading, the equipment item can actually tear itself apart. For example, when a bearing fails in a pump, the shaft can become bent. Remember, the motor is still running, and a bent shaft may cause the spinning impeller to make contact with the pump's involute housing. Also, the pump seals can be destroyed. The net result is a new pump: bearing, shaft, seals, impeller, and housing. Instead of detecting and replacing a failing pump bearing, the entire pump must be rebuilt or replaced.

Clearly, the total cost of downtime because of maintenance and around-the-clock emergency repairs can be staggering. Proactive, predictive, and preventive maintenance are necessary to production operations and facilities to avoid breakdowns and lost production and usage. Part of the process is setting up and using productivity controls for both worker performance and maintenance levels.

Desired levels of maintenance expense should be established for each facility operation against which performance can be measured. This information should be incorporated into an annual operating budget. Also, it is important to determine what percentage of maintenance time is actually productive. Direct and indirect costs should be evaluated.

Maintenance Reports Many reports can be developed to aid measuring and controlling maintenance activities. This may include completed job reports, backlog reports, labor utilization reports, overtime reports, and jobs in progress. **Completed Job Reports** show actual versus estimated performance in terms of cost and time. Large variances can mean poor performance, poor accounting for charges, poor estimating, or poor planning.

Backlog Reports show the amount of work remaining at any point in time. These reports identify whether the organization is current, ahead or behind schedule, and can help determine if the maintenance organization is over- or understaffed. The backlog report can also be used to prioritize maintenance jobs by equipment or area.

Facility managers who have their own maintenance staff can measure backlog in shop hours. Divide the work-hours that remain (unfinished) by the number of work-hours available per day or week in order to determine the days or weeks of work available per employee or department. If there is no backlog, there may be too many people on the maintenance staff. Also, maintenance costs or expenses may be too high. If outside contractors are used for maintenance services, measure backlog in terms of dollars.

The **Overtime Report** shows the number of hours spent for premium time as a percentage of total work-hours. Overtime reports also indicate undesirable levels of breakdown maintenance.

The **Labor Utilization Report** breaks the maintenance payroll hours down into dollars. The report can be prepared from the daily time sheets filled out by each employee or contractor. Activities are identified by code numbers or time sheets. This report shows how much of worker time is spent in necessary but unproductive activities such as delays, travel, waiting for parts, illness, meetings, and so forth. Figure 3.13 is an example of report data generated based on this type of recordkeeping to show the most significant problems of a particular press (in a printing plant) and the nonproductive hours taken up by these problems.

Categories should be developed for collecting maintenance costs. For example, *scheduled repetitive work* such as proactive, predictive, and preventive maintenance activities can be one category. This may include: lighting system maintenance, minor HVAC maintenance, electrical, fire protection, and utility maintenance. *Unscheduled repetitive work* should have its own category. Another category can include *roofing projects, painting, major repairs and renovation,* and *groundskeeping or snow removal. Custodial services* such as cleaning windows and trash removal can be in another category. Parts and supplies should be included for each category. Avoid the frustration of reactive purchasing. Proper scheduling of maintenance will aid in the timely ordering of parts and in the timely maintenance of systems and components to minimize downtime.

Figure 3.14 is a Building Maintenance Checklist that includes these categories.

Use of Ratios to Measure Maintenance Performance Standards should be established for all significant items in the maintenance program, including routine and nonrepetitive maintenance and repair work. Figure 3.15 lists some ratios that you can use to measure maintenance productivity against established standards. Figure 3.16 is pie chart showing the portions of the maintenance budget that might be currently allocated for scheduled and unscheduled, preventive, safety,

Press Report for Month of July–Web Presses

Press 22		Current Month		Year-to-Date	
		Non-Chrgble Hrs.	% of Total	Non-Chrgble Hrs.	% of Total
749	Test Rolls			1 14	0 3
751	Auto Blanket Wash	0 70	0 6	2 97	0 7
752	Slitter Problems	0 34	0 3	3 23	0 8
754	Cocking Plates	3 05	2 7	8 66	2 1
758	Web Breaker Stop			0 46	0 1
759	Ink Drop-Web B	4 49	3 9	13 22	3 2
760	Missed Splice	0 55	0 5	1 13	0 3
761	Change Packing P			1 49	0 4
762	Blanket Washes	0 70	0.6	0 70	0 2
764	Perfing on Number	6 18	5 4	6 18	1 5
765	Auto Bundler	0 85	0.7	1 75	0 4
769	Power Failure			15 61	3 8
770	Idle			50 40	12 3
771	1st Shift Startup	0 15	0 1	22 78	5 6
772	Clean Press/End	5 69	5 0	24 47	6 0
773	Wait-Materials			1 26	0 3
774	Press Repair - M	34 14	29 7	52 11	12 7
775	Press Repair - E	4 50	3 9	14 66	3 6
776	Paster Problems	5 48	4 8	9 53	2 3
777	Folder Problems	2 04	1 8	20 38	5 0
778	Sheeter Problems	0 70	0 6	9 85	2 4
779	Dryer Problems	1 00	0 9	7 61	1 9
780	Slime Hole			0 56	0 1
781	Hair or Calender	0 46	0 4	3 72	0 9
782	Mill Splice			1 40	0 3
787	Picking	4 00	3 5	4 00	1 0
790	Other	0 70	0 6	6 03	1 5
792	Tension Problems			0.81	0 2
793	Mark on Turn			1 36	0 3
794	Unexpl Web Break	4 40	3 8	25 72	6 3
795	Blanket Problems	5 18	4 5	17 10	4 2
796	Rotary Knife Problems			0 22	0 1
797	Dampeners	2 41	2 1	7 70	1 9
798	Ink	0 40	0 3	1 40	0 3
799	Hickey Rollers			0 79	0 2
800	Ink Rollers	8 74	7 6	24 36	6 0
805	Cracked Plt	2 26	2 0	6 64	1 6
806	Blind Plt	0 41	0 4	0 41	0 1
809	Waiting-Plates	1 64	1 4	10 36	2 5
810	Waiting-Prep E	12 93	11 3	26 11	6 4
814	Scratch Plt	0 68	0 6	0 68	0 2
Press 22 Totals		**114.74**	**100.0**	**408.96**	**100.0**

5 Most Significant Problems for Press 22

OP Code		Current Month		OP Code		Year-to-Date	
		Non-Chrgble Hrs.	% of Total			Non-Chrgble Hrs.	% of Total
*774	Press Repair-M	34 11	29 7	*774	Press Repair-M	52 11	12 7
*810	Waiting - Prep E	12 93	11 3	770	Idle	50 40	12 3
800	Ink Rollers	8 74	7 6	*810	Waiting-Prep E	26 11	6 4
764	Perfing	6 18	5 4	794	Unexpl Web Break	25 72	6 3
*772	Clean Press/End	5 69	5 0	*772	Clean Press/End	24 47	6 0

Figure 3.13

Building Maintenance Checklist

Exterior Walls
- ☐ Cracks
- ☐ Rust stains
- ☐ Spalling
- ☐ Corrosion
- ☐ Tuckpointing
- ☐ Painting
- ☐ Loose panels, bricks, trim and cladding

Interior Walls
- ☐ Cracks
- ☐ Settlement
- ☐ Moisture stains
- ☐ Painting

Structural
- ☐ Foundations
- ☐ Columns
- ☐ Roof joists and framing

Roof
- ☐ Debris
- ☐ Drainage and ponding water
- ☐ Exposed roofing felts
- ☐ Ridges and blisters
- ☐ Cracking
- ☐ Flashing
- ☐ Split flashings
  Interior inspection immediately below roof

Life Safety
- ☐ Door closures & hardware; or egress passageways
- ☐ Smoke detectors and alarms
  (Sprinklers and alarms should be checked per the requirements of insurance or regulatory bodies, or manufacturers–by professionals who should present written reports.)
- ☐ Emergency lights
- ☐ Exit lights
- ☐ Fire extinguishers

Grounds
- ☐ Pavement condition (potholes, cracks, alligatoring)
- ☐ Pavement striping
- ☐ Traffic signage
- ☐ Storm sewer system and drainage
- ☐ Fire hydrant condition
- ☐ Snow and ice removal
- ☐ Exterior lighting
- ☐ Landscaping (trees, shrubs, grass)

Conveying Systems
- ☐ Elevators
- ☐ Lift trucks
- ☐ Material handling equipment

Fans–Exhaust, Circulation
- ☐ Lubricate motors
- ☐ Inspect and lubricate bearings
- ☐ Switches and thermostats
- ☐ Belt alignment and wear
- ☐ Clean fan wheels or blades

HVAC
- ☐ Lubricate and check motors, bearings and dampers
- ☐ Adjust dampers, grilles and diffusers
- ☐ Change filters
- ☐ Thermostats
- ☐ Fan belts
- ☐ Clean and check burners, heating elements, and/or coils
- ☐ Zone and safety controls
- ☐ Change filters
- ☐ Refrigerant level
- ☐ Compressor oil and test for acid or moisture content
- ☐ Fan belts
- ☐ Clean coils
- ☐ Condensate pans and drains
- ☐ Zone and safety controls
- ☐ Water treatment

Plumbing
- ☐ Backflow preventers
- ☐ Water pressure
- ☐ Leaky faucets
- ☐ Flush valves and drainage
- ☐ Sewer drainage

Electrical
- ☐ Switchgear
- ☐ Power conditioning equipment
- ☐ Circuits

Process (for manufacturing)
- ☐ Compressed air systems
- ☐ Chilled water systems
- ☐ Water treatment

Figure 3.14

Maintenance Performance Ratios

$$\frac{\text{Total Cost of Maintenance}}{\text{Total Operating Costs}}$$

Maintenance operations include upkeep and repairs to buildings, HVAC, power and lighting, water, gas, compressed air and waste disposal In general, one might use 6%–8% of total operating costs as a benchmark.

$$\frac{\text{Total Cost of Maintenance}}{\text{Total Units Produced}}$$

Maintenance costs should follow the trend of production activity. For example, an increase in the cost of maintenance from 12% to 18% when the number of units produced has decreased may indicate maintenance has been performed unnecessarily

$$\frac{\text{Total Cost of Maintenance}}{\text{Replacement Cost of Machine}}$$

An increase in the cost of maintenance from 8% to 25% in relation to the cost of replacing the machine may indicate a need for major repairs or equipment replacement

$$\frac{\text{Total Cost of Maintenance}}{\text{Total Cost of Labor-Hours Worked}}$$

This ratio indicates the balance between labor costs and the cost of parts and materials

$$\frac{\text{Labor-Hours on Emergency Jobs}}{\text{Total Direct Maintenance Labor-Hours Worked}}$$

This ratio indicates whether too much maintenance is being performed on an emergency basis at premium labor rates Are you managing your maintenance or putting out fires?

$$\frac{\text{Total Maintenance Costs}}{\text{Sales Revenue}}$$

$$\frac{\text{Direct Costs of Breakdown Repairs}}{\text{Total Direct Cost of Maintenance}}$$

Figure 3.15

and overtime maintenance, with a proposed reallocation of funds for replacements.

Other Performance Standards Standards for measuring maintenance performance might, for example, include an acceptable 45-day backlog of work to be maintained at all times. Actual job costs should be within 10% of the estimates. Actual labor productivity targets could be from 95–105% of organization standards for the tasks involved. Unplanned maintenance should represent not more than 10% of total maintenance costs. The standards the organization develops vary with the industry or type of facility, as well as past performance.

Maintenance Contracts Maintenance may be performed by in-house staff, supplemented by outside services; completely in-house; or completely by outside services. As owners and managers of facilities search for ways to reduce operating costs and avoid building-related health and environmental liability, many are placing greater emphasis on proactive, predictive, and preventive maintenance programs. Also, owners and managers can rely on the service expertise of contract maintenance firms to handle their increasingly complex building systems. It may cost more to keep maintenance people on staff than it does to hire them on an hourly basis or retainer.

A maintenance contract should specify what equipment is covered, as well as how often and when (specific dates) it is inspected. Inspections must be made at scheduled intervals. The contracts should state the time frames when the equipment will be available for maintenance. Review the performance of each contract before it is renewed. Keep cost records for each major piece of equipment, and replace the equipment if it has to be repaired often.

Service people should be told to check the complete system rather than solve only the immediate problems. Why just change the fuse without finding what caused the overload? Let the service technician stay to check the complete system for leaks, disconnect switch conditions, contacts or belts, and so forth.

The service technician may discover and eliminate future problems, and should look for the root causes of problems rather than just treating the symptoms. Ask why repeatedly until you are sure the root cause has been identified. Some of the tools that can be employed to determine the root causes of maintenance problems include: the fishbone diagram (cause-and-effect) shown in Chapter 10, and routine recording of data and general statistical analysis. Having determined the causes, you are in a position to select the remedies and preventive maintenance procedures that will provide the best return on investment.

A common reason for inflated machinery repair bills are lost drawings and electrical schematics. Without the drawings, time is wasted either tracing circuits or trying to reach manufacturers for data. It is important to obtain schematics and operating and maintenance manuals for your equipment. Also, maintain a record set of drawings of your building.

List proactive, predictive, and preventive maintenance activities and the frequency at which these activities are performed for each piece of equipment to serve as performance standards and as safety and training aids. The goal is to maintain the exterior and interior of a building for longer asset life at minimum cost.

Maintenance and Indoor Air Quality Indoor air quality problems can result from poor HVAC maintenance. Factors that may contribute to poor indoor air quality include: missing or dirty filters, inoperable equipment, disconnected or uncalibrated controls, and loose fan belts. Also, insufficient water treatment for boilers, cooling towers, and humidification systems can cause fouling that contributes to poor air quality. Additionally, condensate pans and drains can be sources of stagnant water that can release contaminants into the air stream.

Janitorial activities that are not accompanied by adequate ventilation to remove fumes from polishes, waxes, shampoos, spot removers and other odorous products are another contributor to poor air quality. Solvents may be irritating and have toxic effects on occupants.

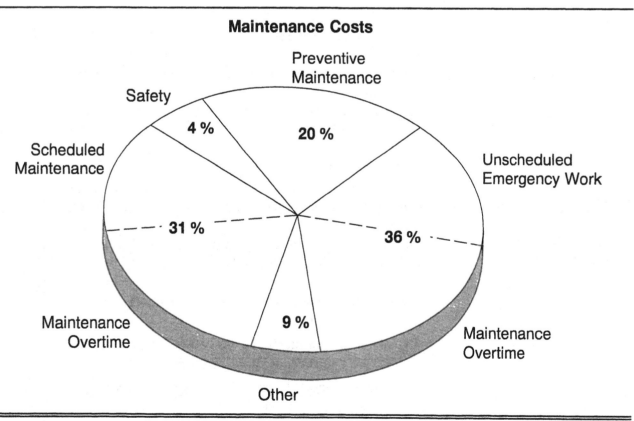

Figure 3.16

Maintenance and Energy Management The quality of maintenance can be a major factor in a facility's energy costs. When predictive maintenance is used as a primary facility management tool, sources of energy waste can be clearly identified and remedied. For example, when infrared detection is used to identify loose connections in electrical panels, those loose connections can be tightened, and energy (measured as heat in this case) can be saved. Remember that predictive maintenance techniques measure excessive or wasted energy.

Total Productive Maintenance Sophisticated maintenance departments in manufacturing environments are implementing total productive maintenance programs to improve overall facility performance, equipment operations, and maintenance processes. Seiichi Nakajima, Vice-Chairman of the Japan Institute of Plant Maintenance, is credited with inventing this innovative system in the mid-1970s, and promoting its use throughout Japan. This concept grew out of the total quality management and employee involvement movements in Japan.

Total Productive Maintenance (TPM) relies on machine operators, maintenance and engineering personnel, and vendors, working as a team to improve the effectiveness of equipment. TPM involves transferring some maintenance tasks to the machine operators. For example, they may be responsible for oiling and greasing the machine. The objective is to increase equipment availability and output without adding personnel or extra shifts. A team composed of both operators and maintenance technicians looks for causes of equipment productivity losses due to set-up times, jams, idling, reduced speed, and other problems, to improve overall equipment effectiveness.

Total productive maintenance encourages machine operators to be curious and advise supervisors about partial failures, unusual sounds, and erratic operation. Total productive maintenance assumes that major breakdowns don't just happen, since breakdowns are typically preceded by intermittent minor problems and repairs, which have been addressed.

Operating personnel are expected to consult warranties and equipment manuals and to use equipment in agreement with manufacturers' recommendations. They are expected to follow a checklist of daily, weekly, monthly, semi-annual, and annual checkups. If equipment is not operated as recommended by the manufacturers, costly damage and downtime will result. If equipment is new and within the original warranty period, improper operation can result in voided warranties.

Goals of total productive maintenance include:
- Zero unplanned equipment downtime
- Zero product defects caused by equipment
- Zero workplace injuries
- Zero waste

- Zero pollution
- Zero loss in equipment efficiency
- Zero equipment breakdowns

The **Overall Equipment Effectiveness** benchmark enables management to determine the equipment improvement potential, develop priorities, and assess the payback for installing a total productive maintenance program.

The formula is:

OEE = Availability x Performance Efficiency x Quality Rate

The world class OEE standard is 85%. Many overall equipment effectiveness calculations turn out to be lower than anticipated. This says that *there is room for improvement.* (See bibliography for a listing of publications on TPM, which cover OEE in detail.)

Figure 3.17a is a graph showing a sample facility's operation's level of maintenance for 22 pieces of equipment, as compared to best practice, or "world class" level. Figure 3.17b shows maintenance cost as a percentage of replacement cost, compared to a best practice level.

Look for ways to take advantage of technology developments to monitor and control the condition of equipment and systems. You may want to use microprocessor controls to monitor and control conditions. Computer and paging systems can be set up to allow the equipment to call managers through hand-held pagers. When equipment is operating outside a control set point, the managers or maintenance technicians can make certain adjustments through remote laptop computers.

Reliability-Centered Maintenance Reliability-centered maintenance is another method for maximizing funds appropriated for maintenance, and for avoiding costly downtime. It focuses on scheduling maintenance activities by calculating the overall system reliability, based on the reliability of components or parts that require maintenance. The reliability of a product or component can be expressed as the probability that it will perform its intended function for a specified time period. When all components must function in order for the equipment or system to operate, the system reliability is the product of the components' reliability. For example, if two components are required for a system to perform, and they each have a reliability of .95, the reliability of the system is .95 x .95 = .90, or 90%. If three parts are required, and they each have a reliability of .95, the system reliability is .95 x .95 x .95 = .86, or 86%. The reliability of a system decreases as the number of components in a system increases. Reliability can also be expressed as the mean time between failures.

Many maintenance organizations fail to implement new facility improvement programs because they are unable to find the time to make the initial investment necessary to learn the program methods, and to collect and analyze the data needed to implement the program. Many of these programs do require a substantial time investment in

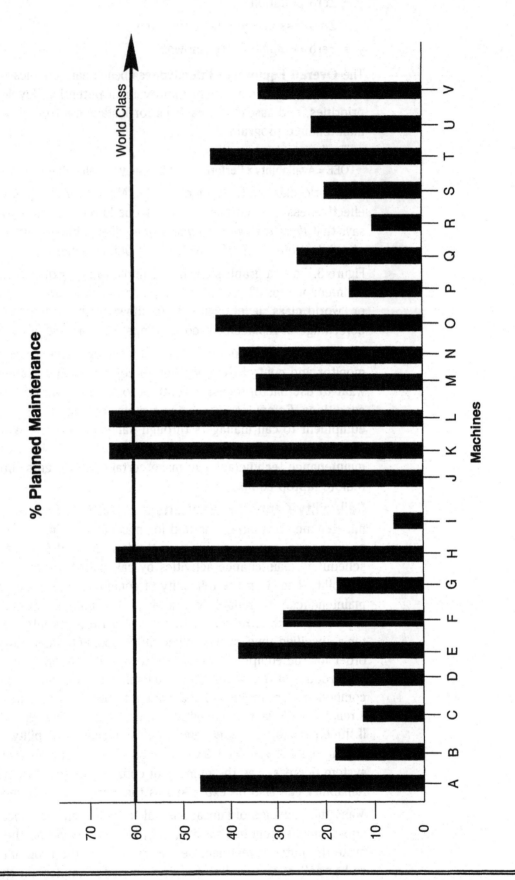

Figure 3.17a

74

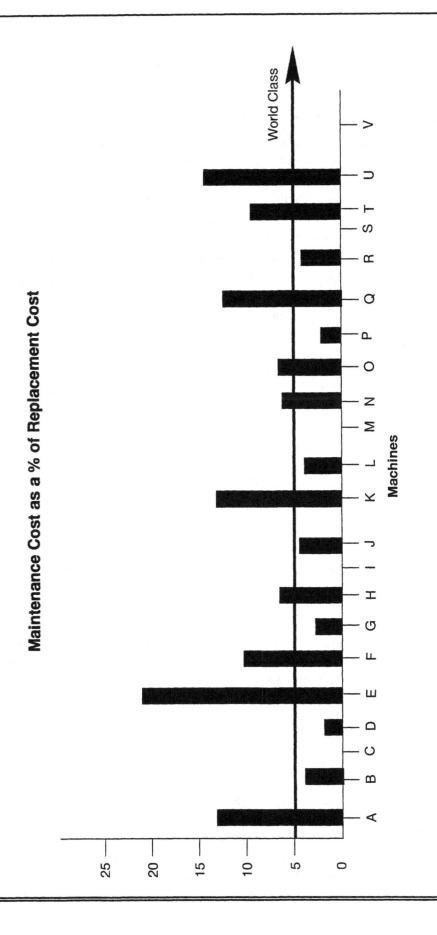

Maintenance Cost as a % of Replacement Cost

Machines

World Class

Figure 3.17b

75

the beginning, but will save time and money over the long run. The cost and time required to implement the programs must be studied carefully to make sure that the benefits outweigh the costs.

Environmental Health and Safety Benchmarks

Indoor Air Quality

What standard should be used to determine air acceptability, safety, and comfort? Do you know if the quality of your indoor air is in compliance with building ventilation codes or within OSHA-permissible exposure levels for contaminants? What is the cost of production inefficiencies and downtime caused by periodic unfavorable indoor air conditions? What are the likely results of temperature, humidity, or air cleanliness problems?

A proper ventilation system must be designed, installed, and operated to remove contaminated air from occupied zones and supply sufficient outside air. The system should be designed to prevent re-entry of exhaust contaminants into spaces, condensation, or growth of microorganisms. Exhausts from most combustion processes contain high concentrations of noxious gases; known carcinogens; and odorless, but lethal carbon monoxide. Outside air inlets should be located away from exhaust outlets, sanitary vents, cooling towers, vehicular exhaust from parking garages, loading docks, and vehicular traffic areas.

The mechanical ventilation system may not bring in a sufficient amount of outside air to make up for exhausted air. When mechanical ventilation systems do not deliver sufficient outside air, drafts may result. Opening doors may become difficult, and excessive infiltration of air through building exterior openings and cracks may result. In more extreme cases of negative pressure, products of combustion containing carbon monoxide from exhaust flues may get sucked back inside the building.

In applications where the chemicals or contaminants being used are below the OSHA threshold limits (and the use is intermittent), and do not lend themselves to local exhaust of the contaminants at the sources, protective respiratory equipment may be sufficient. In manufacturing and office environments, federal, state and local codes regulate building ventilation and indoor air quality requirements.

ASHRAE Standards Because of concerns about indoor air quality and its effect upon building occupants, the American Society of Heating, Refrigerating and Air Conditioning Engineers (ASHRAE) has developed Standard 62-1989, "Ventilation for Acceptable Indoor Air Quality for Buildings." This voluntary standard recommends a minimum of 20 CFM (cubic feet per minute) of outside air per person for a nonsmoking office environment. The old ASHRAE standard was a minimum of 15 CFM per person. If you are considering smoking lounges, ASHRAE recommends an outdoor ventilation rate of 60 CFM per person for those areas.

IAQ

ASHRAE defines acceptable indoor air quality as a condition where there are no known contaminants at harmful concentrations as determined by authorities. The definition includes spaces where a substantial majority (80% or more) of the people exposed do not express dissatisfaction.

New or Altered Building Space Changes in occupant activities and density, chemical usage, operations, and equipment may require a re-evaluation of the existing heating, ventilating, and air conditioning (HVAC) design. This includes determining the HVAC equipment's ability to meet codes and standards. Sometimes equipment, occupancy, and partition walls are changed without consideration for building codes and the impact on the HVAC system. Often ductwork is extended to provide air for new spaces, without verifying whether or not the existing HVAC system is capable of delivering the additional quantity of needed air. This can result in the space having inadequate temperature, humidity, and ventilation.

Existing air handling and distribution systems also may have to be tested to ensure that other areas of the plant are not starved for air. It is not uncommon for contractors to "steal" air from one area of the building to supply air to another space in lieu of adding required mechanical system capacity. Managers should ensure that the quality of air delivered to each space remains within codes and standards. Air distribution systems should be tested, balanced, and adjusted periodically to ensure they are operating in agreement with the design intent and space requirements.

Pollution from Building Materials Out-gassing or off-gassing of pollutants from paints, adhesives, cleaning solutions, synthetic woods and fabrics, carpets, upholstered furnishings, and plastics also may contribute to indoor air pollution. People and their activities conducted indoors, such as cooking, cleaning and smoking, also contribute to the contamination of indoor air. In certain production

---

Indoor Air Quality Ratios

Air Changes
Hour

Clean outside air not only helps building occupants feel good, but it also helps avoid indoor air quality-related health problems For example, it may become necessary to change the air in certain printing-related process areas approximately six times each hour

Cubic Feet Per Minute of Outside Air
Person

Due to concerns about indoor air quality and its effect upon building occupants in office environments, the American Society of Heating, Refrigerating and Air Conditioning Engineers (ASHRAE) recommends in their voluntary Standard 62-1989 a minimum of 20 CFM (Cubic Feet Per Minute) of outside air per person in nonsmoking office environments Higher quantities of outside air may be necessary in process-oriented environments, either to produce adequate dilution of airborne contaminants in the work environment, or to replace contaminated process exhaust air

operations, contaminants in the form of vapors and dusts can cause illness.

Stale air, noxious fumes, and negative air pressures inside a building may signal inadequate ventilation. Symptoms that may suggest poor indoor air quality are: dizziness, burning or watery eyes, shortness of breath, headaches, nausea, general feeling of tiredness, dry throat, chest tightness and skin irritation. The human senses may become dulled after repeated exposure, and symptoms may disappear, leaving an individual unknowingly exposed to a potentially harmful condition. People are aware that the buildings they occupy may be hazardous to their health.

It is recommended strongly that businesses check the quality of their indoor air on a regular, periodic basis. This is a prudent management approach to help guard against potential claims attributed to an alleged indoor air quality problem. It also demonstrates a concern for the well-being of building occupants. Managers cannot disregard claims of discomfort and health problems that are attributed to poor indoor air quality.

According to the "Right to Know" Law enforced by OSHA, businesses must provide workers, EPA and state and local governments with information on types, quantities, and usage of OSHA-listed hazardous chemicals. Material Safety Data Sheets must be obtained from the manufacturer and made available for each operation. The material safety data sheet lists the types of chemicals and relevant health hazard data. The quantity and usage of chemicals must be specified to determine proper material handling procedures, needs for special exhaust and ventilation, and waste disposal requirements.

A review of your plant's material safety data sheet file will indicate potentially hazardous chemicals used in your operations. A periodic industrial hygiene and HVAC system evaluation is strongly recommended to determine levels of potential contaminants, and whether an HVAC system is exhausting properly and supplying sufficient outside air to meet ventilation codes and standards. You need to know that you are within the OSHA-permissible exposure level (PEL) and threshold limits for regulated contaminants. It has been observed that the freshness of indoor air deteriorates as the carbon dioxide level within a space increases. High carbon dioxide levels could indicate a lack of fresh air. You may not win a Workers' Compensation claim for personal damages if a subsequent OSHA inspection reveals that the suspect airborne contaminant exceeded the OSHA-permissible exposure levels.

Other Observations Cooler (but not too cold) temperatures and increased outside air changes may reduce afternoon drowsiness. Greater individual control over temperatures and air circulation should be considered.

In some cases, energy conservation efforts have caused "sick-building syndrome" problems (illnesses directly traceable to the building environment). Super-insulated buildings combined with inoperable window sashes and reduced outside air intake may result in lower energy costs, but, at the same time, may reduce the quality of indoor air. Sometimes monitors can be installed at strategic locations within ventilation systems; they can automatically signal fans to speed up or open outside air dampers wider to increase ventilation rates in accordance with a pre-set contaminant level. This allows owners to optimize ventilation rates for energy conservation.

Solving building-related problems requires a professional, multi-disciplined team approach, and should include personnel representing operations and maintenance, production, design and engineering, and construction. Problems cannot be solved in isolation of other building systems. The impact of a solution on other building systems must be examined.

Engineers should provide flexible HVAC system designs so that modifications can be made easily during a building's lifetime. Changes are inevitable to accommodate both fluctuating occupancy rates and varying uses of the building space.

Thermal Quality

If somebody is too hot or too cold, they probably will not be very productive. Facility management survey results repeatedly identify temperature as the biggest issue in terms of facility use and comfort. ASHRAE Standard 55-1992, "Thermal Environmental Conditions for Human Occupancy," defines an acceptable thermal environment as that condition where 80% or more of the building occupants are comfortable.

It is interesting to note that the number one complaint of building users (according to IFMA's Annual Surveys) is that the building is either too hot or too cold. Although some like it hot and some like it cold, most people will feel comfortable in a space if it meets the following criteria:

- *Space temperature.* Constant space temperature controlled within the range of 72°F to 78°F.

- *Space relative humidity.* A space with relative humidity (rh) controlled between a 40 to 60% range.

- *Air movement.* Continuous air movement with local air velocities of 30–50 feet per minute.

- *Clean air.* Discomfort can be caused by increased levels of airborne contaminants such as volatile organic compounds, suspended particulates, or microbial particles. 20 CFM of outside air in an office environment is generally recommended for dilution ventilation. The ventilation rate will increase with the amount and type of containinants in the air.

Unfortunately, environmental conditions for people and productivity may be different than the requirements for maximizing production equipment and process efficiencies. Certain materials, equipment, and processes are sensitive to changing indoor conditions and require monitoring to prevent problems. For example, relative humidity, temperature, air movement, dust and dirt control are critical in certain manufacturing, health care, computer, and research and development environments. Owners and managers of facilities cannot make occupant health and safety issues secondary to production needs.

Illumination Quality

There are a broad range of lighting requirements in every facility which vary with the activities being performed. Good lighting contributes to productivity, quality of product, staff morale, and conservation of energy. Proper lighting levels with good color rendition and minimum glare are important. Access to natural daylight is uplifting and can be an important energy-saving feature. In a typical office setting, lighting consumes approximately 35–50% of total energy used.

Increasing trends in the use of personal computers and video display terminals, combined with the struggle to minimize energy costs, is changing the way offices are lighted. People are working under lower light levels, which reduce glare and visual stress in a modern office setting.

Light levels should vary according to room function and task location within each room. In the past, lighting professionals generally designed for a uniform level of 75 or 100 footcandles. The IES (Illuminating Engineering Society) of North America (New York, NY) publishes information and standards regarding recommended lighting levels for office and industrial applications. Today, illuminance values in terms of footcandles for various types of activities in interior spaces generally are much less than 100 fc. Typically, 65–75 fc are used for general lighting and 100 fc for task lighting. The trend is to reduce the number of light fixtures and lower the lighting wattage densities. For calculating cooling loads and heat gain from lighting, in the 1970s and 1980s, 3.5–5 watts/S.F. were used. Today, 1.8–2.5 watts/ S.F. are more commonly used. An energy-efficient lighting design can also result in a reduction in the size of air-conditioning equipment.

Fluorescent troffers with energy-efficient electronic ballasts, T8-lamps and deep-cell (three-inch deep) parabolic diffusers are commonly used today. A three-lamp fixture of this type can provide approximately the equivalent illumination as a standard four-lamp troffer with an acrylic lens, while using much less energy. Occupant sensors can also be used to reduce energy consumption.

Acoustical Quality

Erratic noise and continuous high frequency sounds are distracting and stressful to some people and can cause serious and irreparable damage to human hearing. Sound levels should be within OSHA limits. Employee exposure above OSHA's 85 dB (decibel level) requirement based on an 8-hour time-weighted average requires implementation of a hearing conservation program to protect workers. (Further information can be obtained from the *OSHA Guide to Preventing Hearing Loss in the Workplace.*) Studies have also shown that workplace noise can cause an increase in employee errors and impair production, as well as being a safety hazard.

Ergonomics and Safety

The emerging science of ergonomics is based on the concept of adjusting the space or equipment, not the people, to perform a particular operation. Plant layout, material handling, furniture, and production equipment should conform with applicable building codes and regulations, ergonomic standards, the Americans with Disabilities Act, and OSHA guidelines. Unfortunately, many building-related performance problems or complaints fail to reach top management. Sometimes it takes a crisis to communicate to management that improvements are needed.

According to OSHA, in recent years there has been a significant increase in the reported cases of ergonomic disorders in the workplace. The Bureau of Labor Statistics reports that the number of "disorders associated with repeated trauma" has more than tripled since 1984.

Poor ergonomic conditions cause job-related stress, carpal tunnel syndrome, tendonitis, and eye strain. Temporary physical ailments such as muscle strains and backaches, as well as permanent disabilities and fatal accidents, may result from improper layout of facilities or material handling procedures.

Limited aisle spaces and other obstructions may result in workers having to perform tasks in an awkward physical manner, or in a way that puts them at risk to bump into dangerous machinery. The relationships among vertical lift heights, weights of materials, and travel distances must be considered.

Employee costs are rising each year due to rising costs associated with work-related injuries. Medical costs are soaring; there is an increasing tendency toward litigation; and Workers' Compensation rates continue to rise. Worker activities should be studied for hazards due to repetitive tasks, awkward postures, forceful exertions, and vibration.

Project Management

Performance in project management is critical to the success of a facility management organization. Following are some ratios that are helpful for monitoring and evaluating project performance.

$$\frac{\text{Actual Project Cost}}{\text{Budgeted Cost}}$$

$$\frac{\text{Actual Time (weeks, days) to Complete Project}}{\text{Estimated Time (weeks, days) to Complete Project}}$$

$$\frac{\text{Project Management Cost}}{\text{Total Project Cost}}$$

$$\frac{\text{Dollar Value of Change Orders}}{\text{Total Project Cost}}$$

$$\frac{\text{Overhead Cost}}{\text{Total Project Cost}}$$

$$\frac{\text{Budgeted Cost of Work Scheduled}}{\text{Budgeted Cost of Work Performed}}$$

Summary

Competitive success mandates that businesses continually evaluate and benchmark the performance of their facility assets and strategically invest in them to maximize the return to the organization. Systematic evaluation of buildings, development of performance criteria, and measurement of actual performance against planned criteria or standards will lead directly to significant improvements in the quality and productivity of buildings.

Keep management informed about your benchmarking efforts. Publish the results of your benchmarking efforts and use them as an opportunity to gain positive publicity. Figure 3.18 is an example of benchmarking study results distributed to department managers.

The survey results are summarized below.

Things We're Doing Right

- Our mission is clear and well communicated.
- We have eliminated/minimized off-shift staffing.
- We have minimized job classifications. We have two—one of the organizations had forty!
- Our Zero Accident objective reflects a safety attitude and safety performance expectation.
- We have a collaborative management team.
- We utilize decentralized maintenance.
- Our supervisors are multicraft.
- Our communications system includes a computerized network to all offices.
- A well-developed computerized maintenance management system is in place.
- The Zero Roof Leak system implemented is now in year three of five. (Maximizes building capacity; i.e., minimizes customer disruptions.)
- We utilize building automation systems.
- We exhibit a continuous improvement attitude and results.
- We are leaders in environmental friendliness. (Rated Best Environmental Program within the manufacturing sector of the organization.)

Things We Must Do Better

- Develop a system for training mechanics. (Program plan identified.)
- Elevate craftsman role to match responsibility. (Big step completed last June with conversion of mechanics from hourly to salaried nonexempt.)
- Communicate cost savings achieved and dedicate more resources to cost savings activities.
- Implement the use of self-directed work teams.
- Train all employees in "people skills," e.g., collaboration, conflict resolution, and problem solving.
- Speed up maintenance and spare parts storage system implementation to eliminate wasted wait and walk time.
- Strengthen our connection to engineering.

The Business of Facilities Management

We must never forget that for each $1 we defer, reduce, or eliminate, we generate the equivalent of $3 in product sales.

We must continue our basic strategies of:

- Getting rid of what we don't do well.
- Doing what we do well better.
- Trying something new and innovative.
- Doing all three at the same time.

I would appreciate your assessment of the information. Please give me your comments, and we will roll our combined thinking into our department program planning for the next three years.

Figure 3.18

Chapter 4

Value Engineering: Doing More with Less

Our competitive business environment challenges owners and managers to go beyond traditional methods of operation to achieve quality and peak performance cost effectively. As budgets become even more taxed, it is more critical than ever to find new cost efficiencies and streamline requirements. There is mounting pressure on businesses worldwide to become low-cost providers of goods and services. As a result, they cannot afford to deviate from the quest for value in the products and services they buy and sell.

People always know the price of something, but they may not know its value. Your number one job is providing **value** to your business. If your organization does not perceive that you are providing enough value, you will not be there tomorrow.

Businesses have been relatively quick to adopt new management concepts and techniques, such as total quality management, employee empowerment and benchmarking, in an attempt to improve quality and gain a competitive advantage in the marketplace. There is no doubt that application of quality management techniques and competitive benchmarking are necessary in today's world. Unfortunately, with excess capacity and a surplus of available, physical space in many markets, an emphasis on improving quality and efficiency is not enough.

The challenge is to provide a product or service in conformance with customer requirements at a cost that the customer is able and willing to pay. If customers do not recognize the value of a product or service, it will not sell no matter how high its quality or how efficiently it was produced.

Today is the best time in history to focus on providing and obtaining value. This is true for a number of reasons, including global competition, restructuring, downsizing, and technology advancements. People today are more interested in value than ever before. The problem is that top management often does not have a standard method for determining the value of facility projects, products, and services. The purpose of this chapter is to provide facility owners and managers with a customer-oriented, problem-solving method for identifying and eliminating unnecessary facility costs, while improving performance at the same time.

The timing is right for the revival of an old technique called *value engineering*, which has been around since the World War II era. Now and in the future, value management will be an absolute necessity. Value is a fundamental factor in the business success equation. If it is omitted a business is destined to fail. Businesses must search continuously for the best value for the money they spend on projects, products, and services.

There is mounting pressure to improve facilities' performance and simultaneously reduce facility-related costs as a percentage of total operating costs. Without a clear understanding of the actual cost and minimum cost (worth) of providing the required facility functions, senior management cannot measure if it is achieving the greatest value for money spent on facility resources. Customers buy value.

Definition of Value Engineering

Value engineering is a multi-disciplined team approach to identify and remove unnecessary costs while improving quality and customer acceptance based on the analysis of functions. Value engineering studies can be employed as a tool for identification and elimination of unnecessary facility design, operation and maintenance costs, and for the development of facility performance benchmarks. It involves identifying and classifying the functions associated with a product, project, process, or service and allocating costs to the functions. Value engineers define value as the lowest cost method to accomplish the functions required by the customer. The approach is to examine areas that require the greatest expenditures, to challenge the costs and functions, and compare the alternatives on a cost-to-benefit basis. Function analysis is the key differential between value engineering and other management processes and procedures aimed at optimizing the cost/benefit ratio associated with investment decisions.

The method of value engineering, also referred to as *function analysis*, *value analysis*, or *value management*, complements the concepts of modern quality gurus. In a resource-constrained, competitive environment, services must be rendered and projects completed at a reduced cost and within a compressed time period.

Value Engineering Applications in Facility Management

Historically, top management has not given much attention to studying and documenting facility operating and maintenance costs. In addition, senior management did not have a systematic method for determining the value of facility resources and services. Businesses were driven by growing markets rather than cost containment. Today, it is imperative to make value engineering a standard practice in managing facilities. It is a powerful method for identifying opportunities for improvement.

The best place to apply value analysis is where the most facility funds are being expended. Applications for value engineering studies of buildings and facility issues may include:

- To define project functions.
- To define and communicate owner objectives and design criteria for facility development projects.
- To define project scope.
- To establish a budget and set priorities for optimal use of available funds for facility development, operations, and maintenance.
- To establish customer acceptance requirements for evaluating alternative materials, equipment, and designs.
- To determine the feasibility of project ideas.
- To reduce project cost and duration.
- To minimize life cycle costs.
- To determine the optimum ways to comply with code requirements and government regulations.
- To determine procurement procedures for operations and maintenance, as well as construction projects.
- To evaluate methods of construction.
- To analyze management and administrative procedures.
- To prolong equipment life, improve reliability, and reduce operating and maintenance costs.
- To establish maintenance priorities.
- To identify opportunities for reducing waste and energy consumption.
- To reduce cooling loads and select other energy conservation measures.
- To develop target estimates for energy consumption based on functions.
- To reduce building volume and floor area requirements.
- To determine the best use of existing space.
- To select a method for specifying technical requirements for new construction and major retrofit or repair projects.
- To develop procedures and reporting systems for constructing and maintaining facilities.

Facility customer requirements can change over time. Customer standards and expectations for facilities should be reviewed periodically to ensure that facility resources are focused on achieving required functions, and that funds are not spent unnecessarily. Concentrate efforts on activities that add value to the products or services your company offers and strive to eliminate activities which do not add value.

A customer attitude survey (explained in Chapter 1) is conducted during the information-gathering phase of a value engineering study to ascertain customer needs and wants, and to identify opportunities for improvement.

One way to select topics for a value engineering study is to use Pareto's Law of Distribution, which suggests that 20% of items tend to represent 80% of problems or costs. In facility operations, for instance value engineering is suitable for evaluating the heating, ventilating and air conditioning system because HVAC is a large cost center. The HVAC system serving a facility may represent an opportunity for value improvement from a life-cycle cost standpoint. Temperature, humidity air cleanliness, ventilation, zoning, lighting, maintenance and operating parameters can be evaluated during a value engineering study to determine the potential for improvements that can achieve significant cost savings or improve facility performance.

Cost information is essential in any value study. In value engineering, actual costs are allocated to functions evaluated. These costs should include labor, material, equipment, and overhead for each facility, system, component, or service examined. Identification of essential (basic) functions and their costs, plus nonessential (supporting) functions and their costs, permits determination of savings achievable, which leads to the development of the most cost efficient way to accomplish the required functions.

In many cases, there is a difference in the functions of a product, system or service as designed, and the functions actually required, from the customer's perspective. This difference represents a savings potential by eliminating or modifying the unnecessary functions. The worth (value) of a function is determined by comparing the current design or method for performing the function with alternative, lower-cost ways that will also accomplish the same function. The ratio of cost to worth for a function is known as the "value index." In other words, divide the current cost of an existing or proposed design or method by the worth to calculate the cost-to-worth ratio.

$$\text{Value Index} = \frac{\text{Cost}}{\text{Worth}}$$

A ratio greater than 1:1 indicates that there is room for value improvement.

The best value is obtained when the cost of a product, process, or service approaches what it is worth, or when the value index approaches 1.0.

For every facility, there are certain minimum performance criteria that must be met to satisfy the customers' requirements. Value engineering identifies those critical functions and provides cost visibility at the same time. Facility management professionals need to know the costs of products and services in order to manage them and to determine if they are essential in meeting customer needs and wants.

In value engineering, cost is the basis for comparing worth, or value, of alternative methods that will satisfy the customer's required functions. Remember, value represents the least cost necessary to reliably perform the required functions.

The Value Engineering Job Plan

A value engineering study brings together a multi-disciplined team of people who must deal with a problem and have the expertise to identify and solve it. A value engineering study helps people from different departments or functional disciplines who may not normally communicate with one another to work together to solve problems and reach a consensus on what needs to be done. People naturally tend to see things based on their own experience and self-interest. Value engineering forces "islands" within an organization to work together, to gain synergies, and to share information to meet the ultimate task of satisfying customers. If you do that, profits will follow.

A value study team works under the direction of an experienced value specialist who serves primarily as a facilitator. Value studies are accomplished by following a value engineering job plan. The job plan is a set of procedures that the study team follows to review completely the process, product, project, system, or service to ensure that customer requirements are understood fully and met in the form of a cost-effective solution. The method of analyzing value follows the scientific approach to problem-solving and incorporates creativity techniques and teamwork rather than an individual approach to problem solving.

The amount of time devoted to a value engineering study varies, depending on the nature of the problem. According to the value engineering method recommended by S.A.V.E. International, a typical study involves a minimum of 40 hours of team interaction. It is recommended that the value study team be composed of approximately eight people representing a cross-section of the management and technical fields required for the specific areas to be studied. A team with less than four members may not have the resources or talent needed to explore a broad range of alternatives. A very large group may, on the other hand, become difficult to manage and polarize into factions chasing unrelated issues. An HVAC system value study team might consist of a mechanical engineer, mechanical contractor, architect or structural engineer, facility engineer, operating personnel, and maintenance technician, equipment vendor, end user of the space (customer or owner), and value specialist with experience in facility management. Each design decision must be weighed in terms of how it impacts the other disciplines. For example, the specification of a single, large, packaged rooftop HVAC unit in lieu of multiple, smaller, lightweight units may require a more expensive roof frame structure.

The following phases should be included in a value study job plan:

Phase 1: Information
Phase 2: Creativity
Phase 3: Evaluation
Phase 4: Development
Phase 5: Presentation
Phase 6: Implementation

Information Phase

The objective of the information phase is to make sure that the value study team understands the specific problem under study and has collected sufficient information related to it. Problems are best understood after they are broken down into their components. Facility managers need to systematically collect and analyze data concerning customer attitudes, as well as financial and other performance aspects of managing facility-related assets. Information collected during this phase should include space utilization, building system types, capacities, condition, design intent, and operating procedures and parameters, and capital and operating costs.

It should be the duty of facility managers to keep track of facility-related costs so that they can be managed. However, many facility managers and operators do not have detailed historical cost databanks. Some sources of cost data pertaining to facility operations include:

- Monthly utility cost records and operating logs
- Equipment maintenance and repair records and costs
- Design and construction cost breakdown associated with each building system
- Design consultants, contractors, and equipment vendors

Functions Facility managers should break down the total facility expenses, then identify and list the associated functions. A *function* is a performance requirement. It is a generic statement describing what needs to be accomplished without identifying the means for performing the requirement.

Functions are defined using just two words — an active verb and a measurable noun. For example, some functions of an HVAC system may be identified as "cool space," "power fan," "heat air," "filter dirt," "distribute air," "reject heat," "control (or monitor) system." The reason for describing functions using only two words is to force conciseness, improve communication and to break

Information Phase Tasks:
1. Collect and review information
2. Define problem or opportunity
3. Identify and classify functions
4. Develop FAST diagram
5. Identify function costs on FAST diagram
6. Look for any value mismatches

the problem down into manageable segments to facilitate problem solving.

In value analysis terms, the fundamental need of the customer is called the "higher order function," or "task." The task is expressed as a function. The task is the overall purpose or reason why the customer funds the total product, project, or service. For example, the "task" of an air-conditioning system is "condition air." Basic functions of an air-conditioning system are "control room temperature", "filter air," and "distribute air." Once the functions are identified and the task is selected, they can be classified as either *basic* or *supporting*. (See example Function Component Chart in Figure 4.1).

Basic functions are needs that are essential for the product, system, or service to work. The other functions are supporting; these are "wants" that may help increase customer acceptance or marketing of the product, project, or service, but are not *necessary* for it to work. Typical supporting functions for an HVAC system might include: "conserve energy," "minimize maintenance," and "reduce noise." Typical basic maintenance functions include: "inspect equipment," "change filters," "lubricate equipment." Examples of supporting functions are "store parts," and "schedule maintenance." Unnecessary functions are those which are neither required to make the product work or sell, nor required by the customer. After the functions are identified and classified (see Function Component chart), costs are allocated to the functions to help customers understand how their money is being spent.

The FAST Diagram A Function Analysis System Technique (FAST) diagram can be prepared to identify and communicate graphically the classification of functions and their logical relationship to one another, including associated costs. The "task" or "customer-oriented" FAST diagram format as shown in Figures 4.2 and 4.3 was developed by Tom Snodgrass of the University of Wisconsin. It is a variation of the (technical) FAST diagram originally developed by Charles Bytheway in the 1960s. (See Appendix A for an example.) Preparation of the FAST diagram helps team members to reach a consensus on why something is being done and how it is to be executed. Some advantages of FAST diagramming include:

1. It is a tool for analyzing, organizing, and recording functions to stimulate organized thinking and creativity.
2. It is a communication tool.
3. It forces the team to define precisely what is needed or required and thereby understand the task correctly.
4. It aids in locating missing functions by following a logical pattern and helps in checking the validity of functions.
5. It facilitates consensus decision-making.
6. It defines the scope of the subject under study.

Function Component Chart (Information Phase)

Components		Functions	Function Classification	Cost	
			B = Basic S = Supporting	*First Cost*	*Operation and Maintenance Cost*
Air Distribution System:	Fans	Move Air	B	Air Distribution System represents approximately 80% of HVAC system first cost	Air Distribution System represents approximate 30% of operation and maintenance cost
	Ductwork	Distribute Air	S		
	Fan Motors	Drive Fan	B		
	Dampers	Regulate Air	S		
	VAV Boxes	Regulate Air	S		
	Filters	Filter Air	B		
	Grilles, Registers and Diffusers	Direct Air	S		
	Controls	Control Environment	S		
	Detectors	Detect Smoke	S		
Heating System:	Radiators	Transfer Heat	S	Heating System represents approximately 20% of HVAC system first cost	Heating System represents approximate 20% of operation and maintenance cost
	Boilers	Generate Heat	B		
	Piping	Distribute Heat	S		
	Controls	Regulate Heat	S		
		Monitor Heat	S		
	Reheat Coils	Reheat Air	S		
Cooling System:	Chiller and Compressor	Remove Heat	B	Cooling System represents 0% of HVAC first cost (because an adequate system is already in place)	Cooling System represents approximate 50% of operation and maintenance cost
		Compress Refrigerant	B		
	Condenser	Reject Heat	B		
	Evaporator	Absorb Heat	B		
	Piping	Transport Fluids	B		
	Air Handler Chilled Water Coil	Absorb Heat			
	Controls	Regulate Cooling	S		
	Pumps	Remove Water	B		
	Stand-by Package System	Cool Space	S		

Brainstorming List (Creativity Phase)
Function = Distribute Air

Heat recovery system

Variable speed fan

Install separate package units

Single point return

Alternate material for ducts

Return plenum ceiling

Line ductwork

Consolidate ducts

Modify existing ductwork

Divide existing system into two fan systems

Improve diffusers

Balance existing system

Replace existing motors with energy-efficient motors

Prioritize tenant's requirements

Personal fans

Retire steam boiler, install electric reheats to receive electric utility discount rate

Constant volume regulation to switch

Reuse existing fan that now serves switch area

Install window units

Spot cooling

Evaporate air conditioning

Ice system

Open/close windows

Work at home

Figure 4.1

Function Analysis System Technique
Task-Oriented FAST Diagram

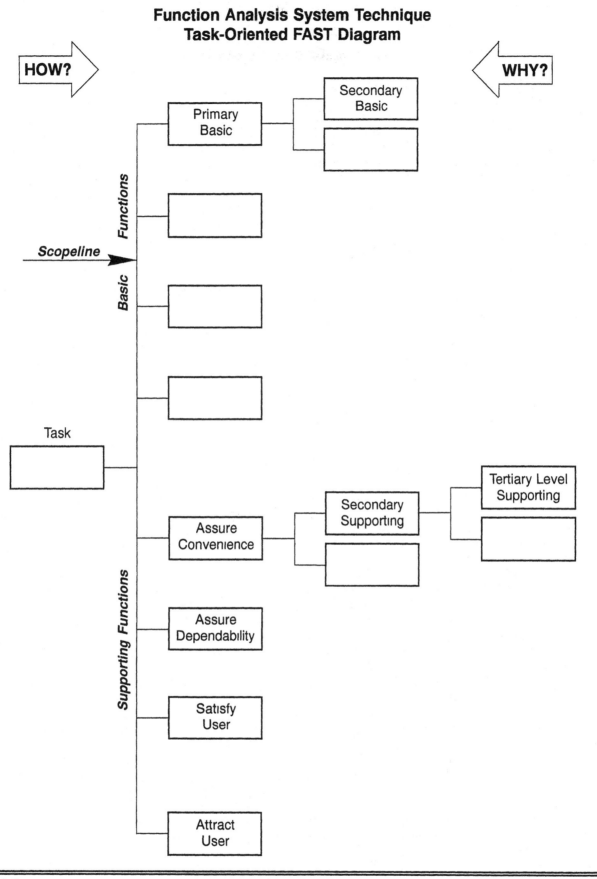

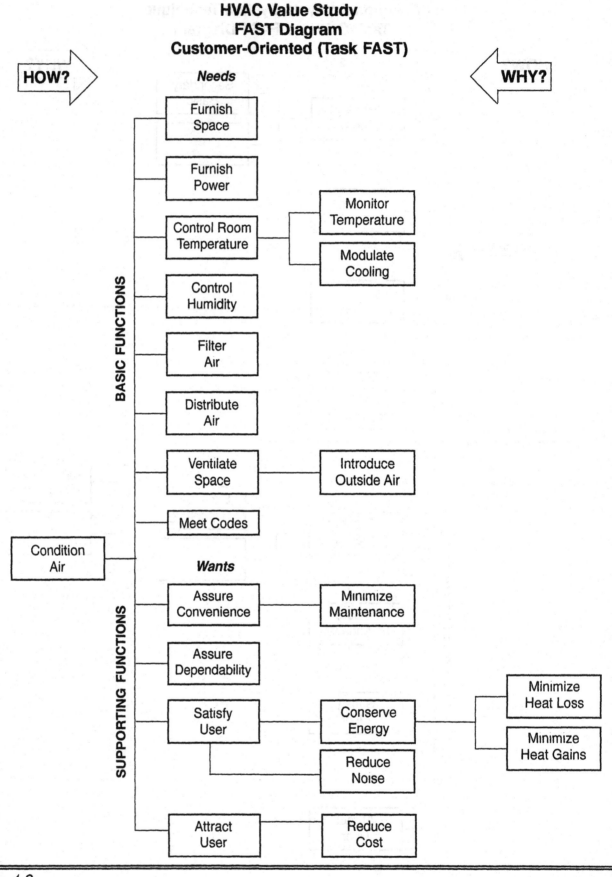

Figure 4.3

94

Questions to ask when evaluating functions of a design, product, process or service include:

1. What is it?
2. What does it cost?
3. What does it do?
4. What must it do?
5. What is the cost of essential functions?

Figure 4.4 is a FAST diagram example focused on a remodeling project to accommodate a smoking lounge. Figure 4.5 summarizes the FAST procedure.

During the information phase, the value study team looks for value mismatches. Value mismatches are high cost functions with low acceptance by the customer. If too little is being spent on a vital service, a mismatch has also occurred. In facility management, if the people doing the actual facilities maintenance or construction work represent 20% of the costs, and the support functions such as clerks and secretaries represent 80% of the operation, there may be a value mismatch. The purpose of identifying value mismatches is to identify opportunities to make positive changes.

Creativity Phase

After the functions have been identified and classified with their allocated costs, the team identifies alternative ways to accomplish the mismatched functions at reduced costs or with improved performance. Brainstorming and other creativity techniques are used to generate a

Creativity Phase Tasks

1. Generate Ideas
2. Record Ideas
3. Suspend Judgment

high number of ideas in a short period of time. (See Figure 4.1) Questions to ask during the creativity phase are:

1. What else will do the job?
2. Can functions or components be eliminated?
3. Can a design, procedure or operation be modified, combined, or simplified to reduce costs?
4. Are aesthetic features necessary?

Ideas should not be criticized during the creativity phase in order to provide a positive environment that encourages sharing of ideas. Following are some typical roadblocks to creativity that may need to be addressed:

• Fear of criticism
• Lack of resources
  • Time
  • Money
  • Personnel
  • Experience
  • Skill

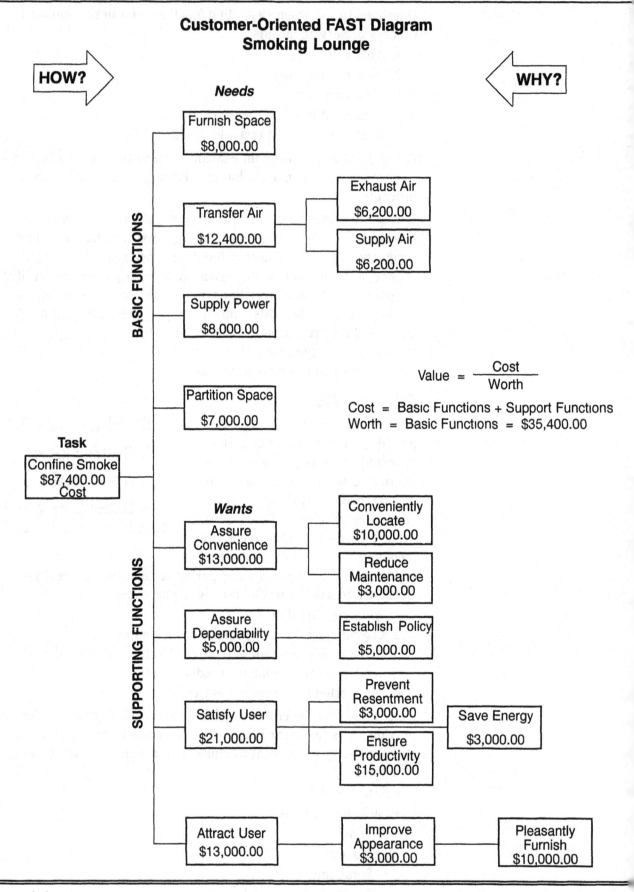

Figure 4.4

FAST Diagramming Procedure

Select Subject, Define Scope of Value Study. Collect Project Study Material.

(For example, for an engineering project, materials could include the owner's design criteria, project budget, building record drawings, utility cost records, maintenance requirements, operations requirements, permit requirements, and other relevant government regulations.)

Identify Functions Using "Verb-Noun" Language.

Functions describe required performance actions without describing specifically the method used to perform the function.

When choosing the words to define the function, make them broad and as open to possibilities as you can.

Write functions on self-stick note pads.

Categorize Functions—Now You Must Separate the Functions into a Single Task, Basic Functions, and Supporting Functions.

Task: The overall need of your customer that is satisfied by the project, product, system, or service. It is the primary reason your customer funds the item under study. People do not buy parts and systems; they buy entire projects to fulfill a desire. It can be difficult to get the team to agree, and it requires a high level of abstract reasoning. Just remember to define the task in terms of what the customer needs.

Basic Functions: Those functions essential to accomplish the task. Think of them as work required to get the job done. Basic functions relate to your customer's fundamental needs and are often called "work functions."

Supporting Functions: Functions that help sell a project, product, system, or service by increasing customer acceptance. Supporting functions meet the emotional, subjective needs of your customer. They are not necessary to accomplish the task, but they do influence your customer's decision to fund your service. Sometimes supporting functions are nonmeasurable, establishing a qualitative statement, such as an aesthetic requirement.

Allocate Costs to Functions

gure 4.5

- Lack of motivation: No rewards
- Conventional thinking habits
- Lack of trust
- Problems with team dynamics
- Mental blocks
- Resistance to change
- Corporate culture
- Highly structured formal setting

Evaluation Phase

The objective of this phase is to criticize and test the alternatives identified in the creativity phase and determine which ones offer the greatest potential for cost savings at the lowest risk. Selected cost-saving alternatives also are refined, cost analyzed, and ranked — from no or low-cost improvements to those requiring capital investment — to determine which ideas are potentially workable. Evaluation phase tasks include:

- Delete impractical cost-saving alternatives generated during the creativity phase. For example, some approaches that may result in facility cost savings may actually create their own set of problems.
- Compare the advantages and disadvantages of alternatives selected for evaluation. Some factors to consider during the evaluation phase may include design cost, construction cost, maintenance and operation cost, constructability, flexibility, estimated payback period, ease of implementation, probability of acceptance by management or the customer, and the ability to monitor and manage cost-savings. Evaluate whether the ideas can be implemented within a reasonable time at a reasonable cost. In the case of the HVAC study shown in Figures 4.1–4.3, these additional criteria were selected: ease of maintenance, appearance, schedule, energy savings, noise reduction, improved comfort, minimal downtime, and system reliability.

Ideas are screened, ranked, and selected for development during the evaluation phase. If a champion cannot be found who is willing to take responsibility for selling and implementing an idea, it is highly likely that the solution will never be realized.

Development Phase

The objective of the development phase is to expand and refine the ideas that survive the evaluation phase. By the end of this phase, one should be able to provide written cost-savings recommendations for consideration by the customer. Procedures for the development phase include:

1. Consult with specialists to verify feasibility of the alternatives and improve on the ideas selected for implementation.
2. Research the alternatives in sufficient depth to prepare specific proposals for implementation.

3. Perform benefits/risk analysis.
4. Detail all results expected, including payback.
5. Anticipate problems relative to implementation.
6. Validate assumptions.
7. Create an implementation plan including cost estimate and schedule.

Presentation Phase

This is the time to present the results of the value study to interested parties, along with a plan for implementing the proposed solutions. Your written report of study results and recommendations should include:

1. Identification of the project
2. Summary of the problem
3. Before and after descriptions
4. Cost of original design
5. Cost estimates of alternatives
6. Technical data to support alternatives
7. Cost estimates for implementing recommended changes
8. Summary of potential savings
9. Acknowledgment of contributions of others
10. Suggested implementation schedule

Implementation Phase

The final phase of a value engineering study is implementation. In this phase, the team ensures that the proposals that have been accepted during the value study are implemented as projects. Figure 4.6 is an example of a value engineering analysis report. See Appendix A for the final report for the HVAC system modifications for the example project shown earlier in this chapter.

Value Engineering Change Proposal

A value engineering change proposal is a means to obtain suggestions from contractors for lowering the cost of a project without impairing performance. It is a proposed change to the contract scope of work, and can be initiated by an owner, designer, or contractor. A value engineering change proposal can be implemented either by an incentive clause or a requirement clause in the contract. A value engineering incentive clause can be incorporated into a contract to provide a monetary incentive for the contractor to suggest voluntary value engineering change proposals. In contrast, the program requirement clause obligates the contractor to perform value studies to the level and scope required by the contract.

Project: Value Analysis of Addition and Alteration to Town Church
Dates of Value Analysis Project: October 27, 1998 to December 15, 1998
Project Hours In a Team Study: 80
Project Staff:

    Value Analysis Study Team Leader

• Architect	• Co-Chairman Building Committee	• Chairman Property Committee
• Mechanical Engineer	for Town Church	for Town Church
• Electrical Engineer	• Co-Chairman Building Committee	• Structural Engineer

    Various Building Committee Members and Specialty Contractors

Project Summary:

A value analysis study was conducted for an addition and alterations project for a town church. After approximately two years of planning and designing, the project was stopped by a vote of the church congregation. The original project was abandoned when a design development estimate was prepared at the 75% drawing completion stage projecting the construction cost to be in the range of $1,200,000 to $1,250,000 rather than $987,000 which was the architect's estimate and budget for the project

Information Phase Observations:

Consultant was retained to perform a value study and to develop possibilities for reviving the project and addressing customer needs. Findings included:

1. Not enough space for the money: the church needed more than the limited classroom, rest room, and kitchen space.
2. Errors and omissions in the Architect/Engineer's design and conceptual cost estimate
    Omitted the need for a new 1200 amp electric service.
    Omitted the need for automatic sprinkler protection of portions of the existing building, in addition to the new building.
    Underdesigned the cooling and ventilation system.
    Omitted patching of pavement for new electric and water service.
    Contained unrealistically low cost estimates for construction trades.
3. Unnecessary aesthetic features
4. Space design did not allow for multi-purpose uses of space.
5. Expensive utilities were incorporated into design to accommodate a Phase II new gymnasium addition.
6. Refurbishment of old gym cost was expensive for a temporary use.

Creativity Phase — Suggestions:

1. Combine design and Construction of Phase I Kitchen and Classroom with rest room space and Phase II New Gymnasium.
2. Design proposed New Gym as a multi-purpose Activity Center that can function as a Gym, Banquet Hall, Theater/Auditorium and Classroom space. Alternatively, leave the old gym in its existing condition temporarily. Existing gym can function as classroom space as is.
3. Construct a larger building.
4. Build a metal building.
5. Conduct classes in homes off-site.
6. Construct new facility in a new location
7. Relocate site of addition to another location on existing property.
8. Abandon addition.

Figure 4.6

Evaluation Phase Summary
- Metal Building: - Not attractive
  - Not acceptable to zoning authorities
  - Not compatible with existing building
- Holding classes in homes not compatible with church education committee and growth goals.

Value Analysis Proposals & Expected Implementation
- Design and construct Phase I and Phase II at the same time as one larger addition in the original location proposed. This would eliminate necessary rework and remobilization costs associated with a second phase. Phase I Kitchen, Classroom, and Rest Room addition contained costly systems for Phase II addition anyway.
- Eliminate unnecessary aesthetic features.
- Change basic function of addition to multi-purpose activity center space in addition to classroom, kitchen, rest rooms, and gym.
- Complete interior build-out of second-floor-level space at a later date. Second-floor-level space would be functional during the interim, but not complete as far as aesthetic features are concerned.

All of the value analysis change proposals recommended above were accepted and implemented successfully by the congregation. The facility was completed and satisfied customer requirements.

RESULTS

Before VE:
Cost: $1,145,000 or $200.00 per square foot
Square Feet: 5700

Functions:
- House Classroom
- House Kitchen
- House Rest Rooms

After VE:
Cost $1,184,000 or $84.00 per square foot
Square Feet: 14,100

Functions:
Increase space over that provided in original plan.
- House Classrooms
- House Kitchen
- House Rest Rooms

House Activity Center, including:
- Gym
- Dining Hall
- Auditorium
- Additional Classrooms

Supporting Functions:
Maximize Flexibility
Reduce Energy
Enhance Appearance

Figure 4.6 cont.

Other Techniques to Lower Cost and Improve Value

Design to budget and *life cycle costing* are other tools value engineers use to optimize value. According to the "design to budget" concept, cost is a parameter in the same sense as technical performance parameters. Designs must be built within estimated cost targets. Decision-makers must also consider the effects on total costs when making project programming decisions.

Facility management involves trade-offs, such as first costs versus operating and maintenance costs. Value engineering helps to determine the most effective way to spend available funds. Value engineering can also be used to develop space allocation standards based on function requirements, rather than the employees' status in an organization.

Value Engineering: Past, Present, and Future

During the World War II era, Larry Miles, a purchasing agent for General Electric Co., was forced to find substitutes for materials that were difficult or impossible to obtain due to the war in order to fulfill manufacturing contracts. He discovered that by specifying performance requirements in terms of functions, rather than the materials or specific parts, General Electric was able to find substitutes that would allow the company to fulfill its contracts at reduced costs, often at improved performance. When Miles started to look at functions, he found that this opened up a new realm of creative possibilities for accomplishing tasks.

Value engineering can be applied to any problem that is expressed in functional terms. The technique also can be used as a tool to improve reliability, maintainability, marketability and performance, but it usually is used to reduce costs.

Although value engineering originated in the United States, it has become more widely used in Japan. The Japanese perform value analysis studies at the beginning of each new product or project life cycle. Korean and Taiwanese firms are also using value engineering. The European Community has even created a Value Analysis Congress.

People in the United States tend to use value engineering as a problem-solving technique only if they are forced to look for ways to reduce costs or if value engineering is a mandatory requirement. It is most often employed reactively instead of proactively when there are project budget overruns, reduced operating and maintenance budgets or organizations are forced to maintain the same programs at reduced costs.

Many federal and state agencies, such as the U.S. Environmental Protection Agency, Department of Defense, Veteran's Administration, and the General Services Administration, require 40-hour value engineering workshops for each new major project. Competition is forcing private industry to take a serious look at the application of value engineering as well. Many organizations now request their suppliers to explore VE applications for improving products and services.

More and more facility management departments are using value engineering to ascertain whether tasks should be outsourced. Before management decides to outsource facility management tasks, it should use value engineering to determine the value of the in-house effort.

Value engineering is a function-oriented team approach to identify and eliminate unnecessary costs. Its strength is derived from analysis of the relationship of function cost and worth from the customer's perspective, cross-disciplinary team problem-solving, identifying value mismatches, using function analysis, applying creativity techniques to generate alternatives, and applying a systematic approach to select and implement the best solutions.

The value engineering methodology uses a scientific step-by-step approach to problem solving that examines the functions associated with a product, process, or service in order to determine precisely what it does, and what it must do to meet marketplace demands. If a so-called value study does not include function analysis, FAST diagramming, creativity techniques, and a systematic multi-disciplinary team approach to problem solving in accordance with a value engineering job plan, it really is not true value analysis.

Organizations must thoroughly research functional, technical performance, and financial requirements before committing major capital and operating funds for facility development & maintenance projects. Value engineering provides the tool to accomplish this goal.

Part II

Project-by-Project Improvement

Chapter 5

The Project Management Process

Customer attitude surveys, SWOT analyses, life cycle cost analysis, benchmarking, and value engineering are management methods which determine the changes that are needed to support an organization's mission and objectives. Projects are then established to carry out the changes required. Project management is the vital methodology for implementing desired changes.

The purpose of project management is to reduce risk factors and accomplish desired changes as effectively and efficiently as possible. The project management system is a means for providing coordinated management of the individual projects. To successfully manage projects, it is necessary to organize the work effort into a set of tasks and assign responsibility to an individual or small group to do the job. Some benefits of the project management approach include: efficient allocation and control of limited resources, shorter overall project cycle times, minimum cost, and improved quality and performance.

Definition of Project Management

Project management can be formally defined as the systematic application of specialized management techniques by an individual or small group who has the responsibility for meeting project objectives. Project management is based on the following concepts:

1. **Management of the project risks**, from concept to completion of the project, by a project manager. To be effective, the project manager should have the expertise to manage projects and be given the resources, responsibility, and authority to plan and control the project.
2. **Use of specialized project control methods**.
   Examples include value engineering, critical path method scheduling, cost/benefit analysis, contract procurement and administration methods, and conflict management skills.
3. **Performance of the project scope of work** by a multi-disciplinary team of specialists directed by the project manager.

Characteristics of Projects

The classic definition of a project is a *unique undertaking with a defined scope, cost, and start and finish date*. Examples of projects include: building or remodeling a facility; conducting a feasibility study; introducing a new program into an organization; reorganizing a department; or modifying present facilities to meet building codes, environmental requirements, or new production standards.

A project is very different from a set of ongoing business activities such as daily accounting, mass production operations, and routine maintenance. Many projects typically involve risk and uncertainty because of the uniqueness of the undertaking, complexity, use of limited resources, time and cost constraints.

Creating a Project-Driven Facility Management Organization

Increasing competition and shrinking profit margins are forcing businesses to find ways to obtain better control and use of facility assets with smaller staffs and operating budgets. To accomplish this, organizations establish projects and structure the facility management organization to facilitate the planning and management of projects.

Facility management is well-suited to a project management approach. Much of the work requires a high level of coordination and cooperation with other functional departments of an organization. Many facility management activities are unique, one-time undertakings. Workloads and required staff skills are unpredictable and fluctuate significantly. It is not economical to retain permanent staff levels to meet peak demands during indefinite slow periods.

As an extension to the project-driven facility management organization, many facility managers have moved toward reducing staff to key personnel and filling other positions with outside contractors. Outsourcing and out-tasking of certain facility-related services are used to level out staff workload fluctuations and to obtain expertise unavailable from permanent in-house staff members.

Instead of staffing facilities departments to meet peak workloads, departments are being staffed for normal workloads, and outside contractors are retained for periods of heavy workloads. Specialized project management skills and techniques are required to plan and control projects, and to develop better technical and operational specifications, bid forms, and contracts for both internal and outsourced work. Outsourcing has become a viable organizational strategy for maintaining or improving performance at a reduced cost. Outsourcing becomes a "surge protection" for dealing with periodic surges in the personnel needed to achieve corporate objectives.

Outsourcing and Out-Tasking

Outsourcing, as opposed to out-tasking, involves turning over the complete management and decision-making authority of an operation (e.g., maintenance) to somebody outside the organization. The management contract for an outsourced operation may be for a period of three or four years. Out-tasking, on the other hand, involves keeping the overall technical competency and decision-making authority in-house and using outside specialty services, on a temporary basis, to supplement in-house capabilities (e.g., manpower, technical expertise).

Some people believe that in-house business support services (non-core business competencies or nonrevenue-producing services) are monopolistic and have little incentive to improve efficiency and effectiveness of their operations. Therefore, non-core business services, such as building operations, janitorial and maintenance, real estate, facilities engineering and office support, should be performed entirely by outside contractors.

This kind of thinking may be overly simplistic. It may seem like common sense, but needs to be thought through. While we are all looking for easy answers to our problems, company objectives, politics, and philosophy cannot be overlooked. Facility management is a sophisticated business and typically represents a significant amount of a company's operating expenses. You must think of yourself as the CEO of a facility management business. For example, if you are a facility manager with a $3.5 million annual facilities budget and a staff of 25 people, essentially you are the CEO of that organization. Just like the corporate CEO, you are operating in a competitive environment with constant pressure to reduce costs and improve performance. The quality, value, efficiency, and effectiveness of the services you provide will always be compared with what is available in the marketplace. Your organization must be able to compete with the *best in class* facility service providers. This is why benchmarking is important. Running a business is like playing chess. If you play with a "checkers" philosophy, you will never become a chess master. Your operation should meet or exceed best practices consistent with the objectives and culture of your company.

How do you determine whether something should be outsourced or out-tasked? If it can be done faster, cheaper, better, or safer by someone else, then the answer is to outsource or out-task. There are some specialty services that should not be outsourced or out-tasked. For example, it may not be feasible to outsource maintenance tasks at a nuclear weapons manufacturing plant because of national security and safety reasons. If manufacturing is the core competency of your company, you probably will not want to outsource the maintenance of certain equipment and systems that could result in a liability lawsuit or loss of market share if a defective product is introduced into the marketplace.

Deciding to Outsource or Out-Task

List the services provided by your organization and determine how much they cost and how long they take to perform. Use Pareto's Law of Distribution to identify the twenty percent of work items that comprise eighty percent of the cost (dollars). Perform external benchmarking (a comparison of your practices, costs, and performance results with similar businesses outside your organization) as a reality check. You must look at external benchmarking to determine the best practices and where you want to be. See what other companies are paying for comparable services. As a simple illustration, you may discover that someone is paying much less for carpet cleaning on a square foot basis than you are. Remember, when you go through the benchmarking process, you may discover a better practice.

Benchmarking based on internal standards of performance alone can lead to fatal analysis. Most of us believe that we have the best kids. Our work is where we spend a lot of time, and most of us are proud of what we do and what we produce. Looking at external practices can be a humbling and enlightening experience. You must keep up with the best to survive and compete.

The next step is to perform a value assessment on the highest cost items. This is a better approach than making a blanket decision to outsource everything.

On the surface, there appears to be very little risk associated with outsourcing building operation and maintenance functions. In many cases, the key people who have been performing the work as direct employees of the company will often stay on as employees of the outsourcing management company. The difference is that they now are working for someone else under a three- or four-year contract. They may or may not have the best interests of your company in mind. Like anybody else, they may be inclined to be loyal to the person or entity whom they believe will have the most positive impact on their future career development. The outside service provider is typically awarded a contract based on the premise that they can perform the same service faster, cheaper, and better. It is possible that the former employees who now work for the outside service provider may be reinvigorated because they now see a potential to reach new career heights with the outsourcing company whose core competency and service matches their own skills and career interests.

In such situations, personnel can feel a sense of security, knowing that they will continue their same job working for another employer, at least for the term of the outsource contract. They have an incentive to perform and keep their former employer (new customer) happy so that the contract will be renewed when it comes due and provide employment for a longer period of time.

Suppose the outsource provider says that it will guarantee a ten percent cost reduction over the current cost of running a facility operation. In order to reduce the cost, the outsource provider may eliminate senior facility management personnel and other staff who are overpaid according to the market or not carrying their weight.

The outside service provider is free to compensate personnel based on performance and not seniority.

Another issue is that the outsource management company may not have the same incentive as a company employee to communicate the additional work required to maintain, repair, or replace equipment and systems if its main objective is to reduce costs during the life of the outsourcing contract. At the end of a three-year outsource contract, your company may have saved three hundred thousand dollars, but have a one million dollar problem.

Also remember that when you outsource everything, you contract away your knowledge and the power that goes with it to an outsider who is motivated to perform only those services that it is contractually obligated to provide. The tasks, maintenance priorities, frequency of maintenance, decisions about methods, parts to use, and task requirements become the responsibility of the outside service provider. Continuity and knowledge may be lost forever if the outside contractor fails to perform and its contract is not renewed.

Ironically, a well-run company that keeps its equipment operating like new and follows best practices may reduce life cycle costs without sacrificing performance by outsourcing. In companies with equipment in various states of disrepair that have not been operated and maintained properly, it will be difficult to achieve short-run savings. It may take a substantial investment by the outsourcing company to restore the equipment to original design standards.

A major problem is that many companies cannot find the time and do not have the resources to do in-depth benchmarking. If you are using 95% of your resources to carry out routine operations, then you can only devote 5% of your time to continuous improvement efforts. Continuous improvement activities too often get dragged out because of lack of time.

Most project failures can be traced back to limited or improper management. Before a project is started, it is necessary to assess the strengths and limitations of in-house staff members. Some tasks can be handled best by outside professionals with specific technical expertise.

There must be a cost-effective balance between use of in-house professionals and services provided from the outside professional. Cost, experience, technical and personnel resources, response time, and quality of services are some factors to consider. Qualifications of consultants and contractors are discussed in Chapter 10.

Project-driven facility management departments are becoming more marketing-oriented and actively seek projects to manage within their own organizations rather than waiting for projects to be brought to them. This approach also assists them in planning and scheduling project requirements to maximize efficiency of scarce resources. In larger companies, some customer-oriented facility departments have even established a position for an account representative. This representative's job is to interact with customers, ascertain their needs, identify future projects, and then schedule and coordinate the required resources. The principle of this facility management reorganization is to proactively address customer needs while allocating facility management resources more efficiently.

Key Aspects of Project Success

There are three main areas to focus on in planning a project. The first is the choice of project delivery method, which will depend on the type, complexity, and time frame of the project. Second is organizational issues related to your internal project team. Third is budget, cost, and schedule considerations.

Project Organization Structures

The organization structure is an important factor in determining project success or failure. A common cause of poor individual and group performance is inadequate organization. The organization structure represents the planner's strategy for organizing functions and people to achieve the goals of a business concern in the most effective manner. Consequently, if the coordination of functions and units is awkward or cumbersome, it will inhibit or make difficult productive and effective work. When an organization does not have a system for accomplishing desired results, it is disorganized. In today's dynamic business environment managers must be able to design flexible organizations that can readily adapt to changing customer needs and expectations.

Critical Choices and Elements for Project Success

I. **Project Delivery Approach (Process)**
   (Consideration and selection of the appropriate project delivery method)
   A Designer and General Contractor
   B Design/Build
   C Construction Management
      1. Contractor CM
      2 Extended Services CM
      3 Agency CM

II. **Organizational Issues**
   A Competency of project team
   B Camaraderie between team members
   C. Communication (Feedback)
   D Competency of project manager
   E Top management support
   F Responsiveness to clients
   G Sufficient resource allocation
   H. Technical tasks and procedures
      1. Client acceptance reviews
      2 Technical team reviews
      3 Procedures and controls
         a Quality
         b Documentation
         c. Communication and flow of information

III. **Programming**
   A. Value engineering and life cycle cost analysis
   B. Definition and control of scope of work
   C Estimating and cost control
   D Schedule control

Organizational charts can be useful to illustrate the formal relationships and lines of communication between work groups in an organization. However, they do not identify the decisions the individual work groups are authorized to make. A decision chart such as that shown in Figure 5.1 is a helpful tool for resolving conflicts between groups of people operating within a complex organizational structure. Different work groups may share responsibility, but approach resolution of issues from their own perspectives. Try to minimize shared decisions where possible in order to reduce infighting within organizations.

There are three major organizational models: *functional organization*, *pure project organization*, and *matrix organization*. Just about every organization has all of these structures with numerous variations.

Functional Organizations

The functional organization is the traditional organization with separate departments or work groups, each having members with similar skills, disciplines, or functions. The organizing principle behind the functional organization is *knowledge*. The theory is that it is easier to manage specialists if they are grouped together and directed by someone who possesses similar expertise and experience.

The functional organization is appropriate for stable environments where decisions and work tasks do not require much interaction or coordination between different departments or specialized work groups.

The functional organization tends to be hierarchical with many layers of management (e.g., director level, division level, department level, section level). Communication channels are vertical and well established. Unfortunately, when information travels through the hierarchies, it tends to be distorted and rearranged. People in different areas of specialization, in a sense, speak different languages and naturally approach issues from their own perspectives. When there is a problem in the organization, they often point their finger at other departments. The horizontal flow of information between departments in a functional organization is often hampered because each department tends to promote the interests and values of its members.

While it is foolish to attempt to organize without the aid of an organizational chart, a chart alone is limited in the practical information it can provide and can oversimplify the interrelationships among members of the organization. Since the chart is a formal approach to organization, it cannot show informal working relationships. Despite its weaknesses, the chart is valuable in that it puts a structure in writing, thereby lessening the tendency to pass the buck. It also helps people know how their jobs fit into the total picture, an essential element in maximizing productivity.

Decision Chart for a Manufacturing Organization's Facilities and Engineering Departments

Decisions	Engineering	Facilities	Maintenance
Daily Maintenance Operational Issues	I	I	R
Daily Facilities Operational Issues	I	R	C
Manufacturing Equipment Modification Issues	R	C	C
Distribution Equipment Modification Issues	R	C	C
Facilities Equipment Modification Issues	I	R	C
Long-Term Manufacturing Strategy Issues	R	C	C
Long-Term Distribution Strategy Issues	R	C	C
Long-Term Facility Strategy Issues	S	S	C
Manufacturing Capital Planning Issues	R	C	C
Distribution Capital Planning Issues	R	C	I
Facilities Capital Planning Issues	S	S	C
Manufacturing Equipment Maintenance Issues	R	I	C
Distribution Equipment Maintenance Issues	R		C
Facilities Equipment Maintenance Issues	I	R	C
Manufacturing Equipment Modification Issues	R	C	C
Distribution Equipment Modification Issues	R		C
Facilities Equipment Modification Issues	I	R	C

S = Shared Decision
I = Informed
R = Responsible For
C = Consulted

Figure 5.1

Figure 5.2 is a typical functional organization chart. Examples include the marketing, administration, production, engineering, and facility management departments. Although responsibility for accomplishing project objectives may belong primarily to one department, each functional division tends to be oriented toward the activities of its own function, and functional demands often take precedence over project needs. Lack of coordination among project activities is typical as responsibility for the project shifts from department to department throughout the various project phases. This results in communication breakdowns, buck-passing, and varying commitment levels from each department. The project may be of critical importance to one department, but considered minor to another, thereby impairing its successful completion.

Potential Problems Following are two examples of the effects of a project shifting departments. The division accounting department of a large corporation purchased new project management job-costing software, without input from the facilities and services engineering department. When the accounting department got bogged down with installation and implementation of the software system, it transferred responsibility for implementation to the facilities and services engineering department in order to contain costs. Another example is a facilities department receiving funds for a major construction project, then discovering that it had failed to include some essential items in the project budget. As often occurs, a bad budget led to a bad decision. The facilities department decided to compromise on the design criteria for the climate control system, which created some headaches for the production department after the project was completed.

Under the functional organization, the project is divided into segments and assigned to relevant functional departments. Project activities are coordinated by functional departments and upper layers of management, rather than by a designated project manager. The functional organization is traditionally a top-down directive approach to management, which discourages two-way communication, teamwork, empowerment, cross-functional systems thinking, and creativity. Leading management consultants believe all of these characteristics are essential to continuous improvement of quality. Bottlenecks and delays occur when there are too many projects in a functional organization with many layers of management. Decisions and resolution of conflicts are pushed up to the senior managers who are overloaded and unable to make decisions in a timely manner. Departmental barriers must be broken down to generate breakthrough improvements in quality and performance.

Numerous businesses have reduced layers of management and cut staff to minimum requirements. Some larger organizations have switched from a functional to a matrix organizational form in order to decrease in-house staffing requirements and avoid communication barriers and other problems prevalent in functionally organized structures.

Pure Project Organization

The pure project organization (also called the *projectized organization*) structure allows the project manager maximum control over resources necessary to accomplish the project objectives. A project manager is assigned to oversee the project and has primary responsibility and authority to achieve successful completion of the project. Functional personnel necessary to complete the project are assigned on a full-time basis to provide technical expertise to the project manager until it is deemed that their services are no longer required. The project becomes a self-contained unit with its own technical and administrative staff. Project team members may defer to functional managers only for technical advice. The project team remains linked to the parent organization through periodic progress reports. Under the pure project organization, the project manager is like a president or CEO of a firm whose sole reason for existence is to achieve the project objectives and meet the customer's expectations.

Typical Functional Organization Model

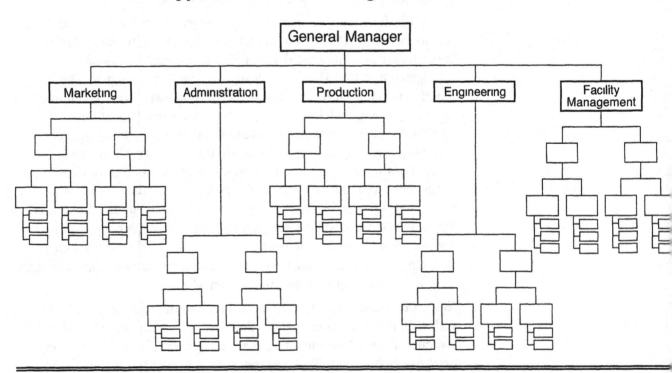

Figure 5.2

The choice of the appropriate organizational structure for a project ultimately depends on the project objectives, size and complexity of the project, work necessary to accomplish the objectives, the technology to be employed, environmental factors, resources available inside and outside the parent organization, and the customer's needs. The best application for the pure project management organization is a large, complex project with a big budget and long project life cycle, such as construction of a major corporate headquarters facility that would require over a one-year time commitment. The pure project management structure also minimizes conflicts arising from activities that cross functional organization boundaries, because each functional unit assigns a full-time representative to the project. The pure project organization may be composed of part-time or full-time members. Figure 5.3 is an example of a pure project organizational chart, in this case for a manufacturing company's new construction project.

Matrix Organization

A matrix structure attempts to combine some of the advantages of the functional organization and the project-driven organizational structures. Where cross-functional problem-solving and teamwork are required to achieve rapid change throughout an organization, a matrix organizational structure is preferred.

The matrix organization permits the management of numerous projects simultaneously without the cost of adding full-time project personnel. In matrix-organized projects, a project manager is appointed from within the organization, given total authority over project activities and flexibility to acquire needed resources for projects from within or outside the parent organization (subject to time, cost, and performance constraints). Project management personnel are specifically assigned to manage selected projects from concept to completion. Within matrix organizations, the project manager controls administrative decisions (project scope, cost, time and quality) and functional heads control technical decisions. The matrix organization enables the project manager to temporarily draw from the technical expertise within the functional departments. (See Figure 5.4, a matrix organizational chart.)

People working within a matrix organization frequently have two bosses: the functional department to which they are permanently assigned, and the project manager for whom they temporarily work in order to accomplish cross-functional objectives of the overall business enterprise. In these situations, personnel may be caught in the middle of conflicts between the two supervisors or customers. This can be a difficult role for some people.

A matrix organization supports maximum efficiency in the utilization of scarce resources. Theoretically, project costs are minimized because key people can be shared. Job security is strengthened because at the project's end, project personnel still have a home

Pure Project Organizational Model for a Manufacturing Organization's Construction Project

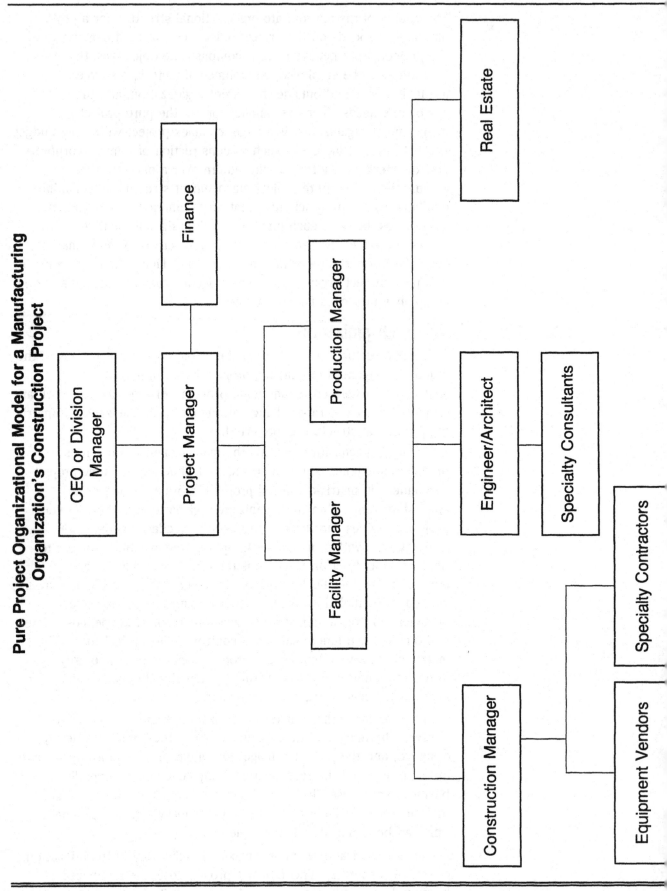

Figure 5.3

Matrix Organization Model

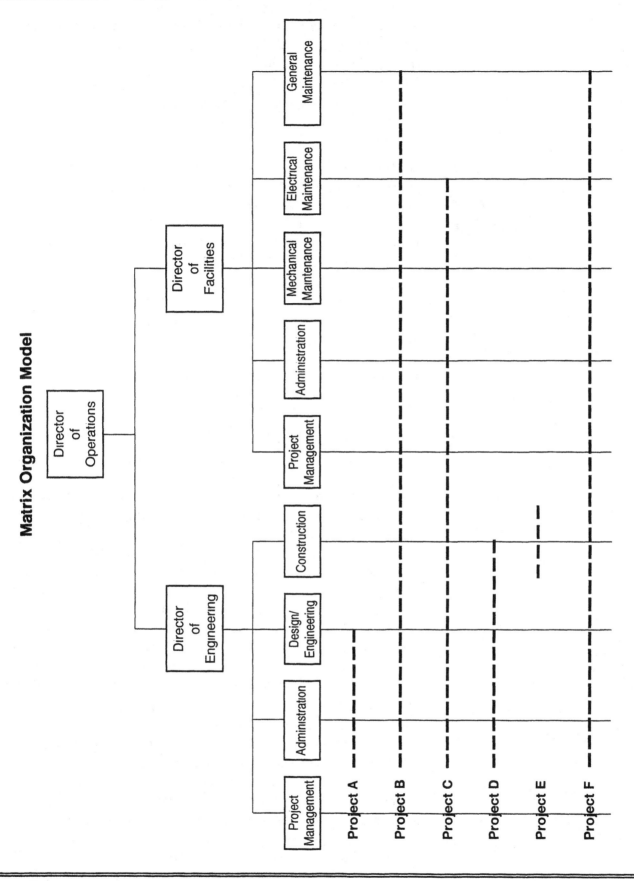

Figure 5.4

within their functional department. The downside of a matrix organization is the tendency for power struggles and infighting between the project and the functional department managers over who has the greater authority regarding use of personnel resources. The project managers must integrate various department resources to promote the interests of their projects, often begging, borrowing, and otherwise negotiating for what they need.

A Hybrid Organization Structure

Functional, matrix, and project organization structure designs are based on a specific purpose, and have been traditionally utilized over the years. However, there are many instances where any one of these specific models alone will not meet the needs of an organization's operational environment. In such instances a hybrid of the three traditional structures may be appropriate.

One of the first steps in designing a new organization structure is to analyze the environment within which this structure will function, its purpose, and the customers the organization will service. In just about any situation an organization will provide some type of service, either tangible or intangible. A good tool that can be used to evaluate the environment and define the purpose and service of an organization is a one-on-one interview with each prospective client. We all have clients, both independent business units and service departments within large organizations.

Begin by identifying your organization's customers. Approach each customer with the idea that you are in the process of designing an organization to service their department needs, and you are interested in learning from them what those needs are and how they feel you can help them. This discussion will accomplish several objectives: It will help you develop a positive relationship between your organization and your customers; it will give you some important information about your customers' needs; and it will facilitate your customers' "buying into" your new organization and its value. Once you have interviewed your customer base, document all this information in an organized manner so that you can use it to evaluate the situation and design the organization. You will also need to perform the same task with potential members of the organization you are trying to design.

The task now is to somehow take all this information and develop an organizational structure that fits the situation. This is not an easy task. In many cases, you will find that a hybrid of the three traditional organizations mentioned above will work better than any one of these structures. Be careful not to fall into the trap of trying to fit your need to one of the traditional structures. A crucial requirement of your organization is that it fits the needs of your customers (which is the purpose for the structure), and that the structure is built on the strength of the individuals who will be part of this new organization. As you begin brainstorming to create the structure of your organization, use a spreadsheet to organize your thoughts. A generic example is shown in Figure 5.5. Lay out your customer needs (projects you must support) in the left-hand column from the top down, and your facilities department functions across the top row starting in the second column. Although this tool will not necessarily give you an instant picture of the appropriate organizational structure, it will enable you to look at the needs of the organization in an organized manner. After several iterations of this exercise and discussions with your customers (as well as potential members of the organization you are trying to design), you will begin to develop a potential organizational structure in your mind. You now can start drawing your organization chart. (Software such as "OrgPlus" from Broderbund Banner Blue Division can be helpful for this task.) After several iterations, discussions and modifications, your organization structure will begin to acquire a shape.

Figures 5.6 and 5.7 show organization structures of two existing service groups within a manufacturing organization. The objectives of this example are to design an organization that fulfills the customer needs, builds on the strengths of the individuals within the organization, and provides potential for people to develop. Interviews with the current and potential clients, and with members of the old groups resulted in the following feedback:

- Very little focus on customer service
- Poor communication causing confusion about who is responsible for what project
- Poor reporting of project status to customers
- Lack of succession planning

The diagram shown in Figure 5.5 is then created based on the overall customer's (functional departments) and the support areas' needs. After evaluation of this information and several iterations of organizational structure designs, the structure shown in Figure 5.8 is developed. Everyone is involved in this exercise, both clients and group members. The benefits of the new organization structure are as follows:

- Provides the opportunity for succession planning. By placing the right people in the right positions, you can make the most of their capabilities, giving them opportunities to grow and move up in the organization.
- Provides the opportunity for overall employee development. Employees may be rotated to other positions where they will benefit from cross-training and exposure to other aspects of the organization.
- Provides a focus on customer service. This is the purpose of developing the project operations group. By having a project manager (e.g., A, B, C, in Figure 5.8) dedicated to a specific functional unit, e.g., printing or packaging, there is better service and better focus to resolve problems and meet customers' needs.

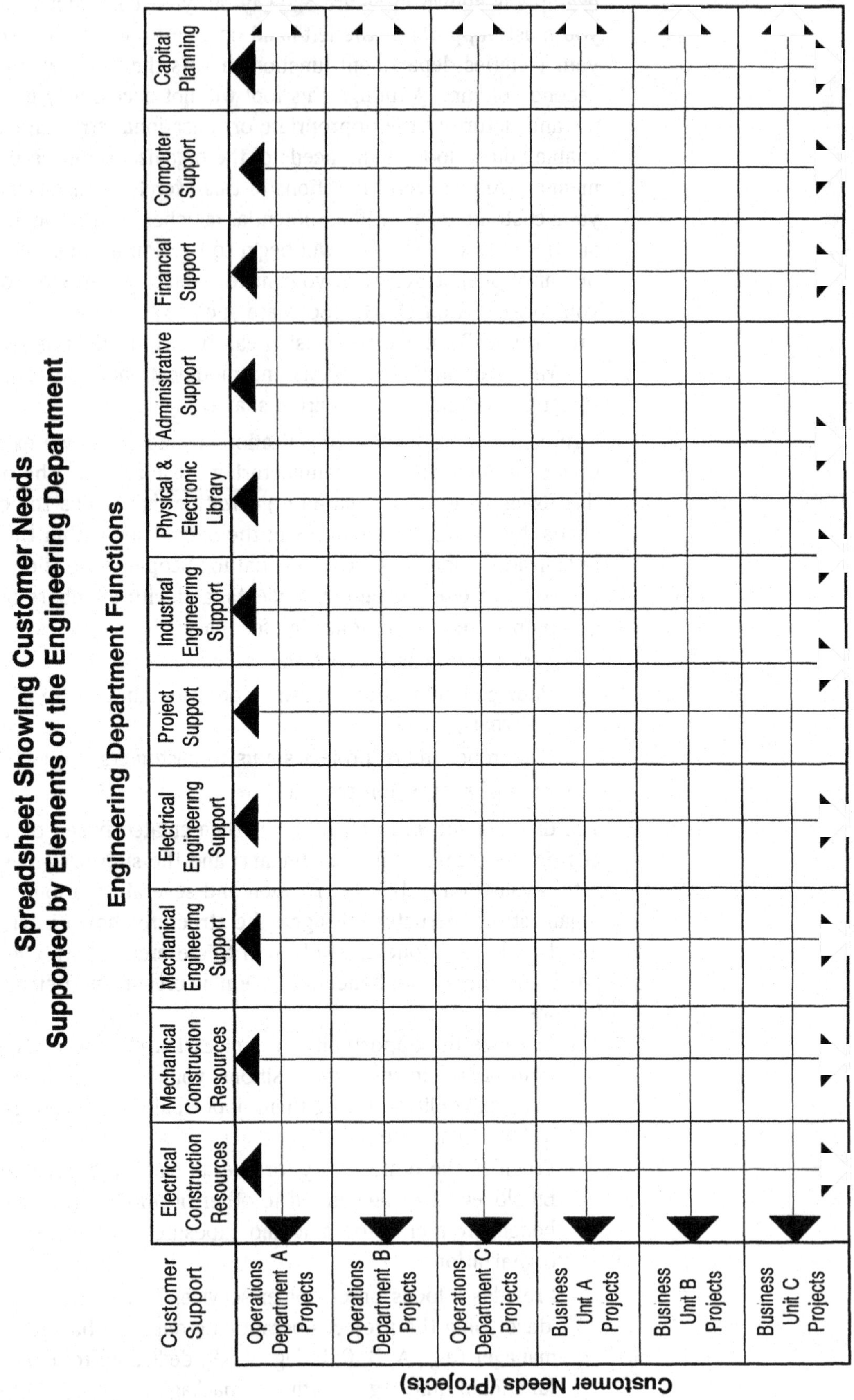
Spreadsheet Showing Customer Needs
Supported by Elements of the Engineering Department
Engineering Department Functions
Customer Support
Electrical Construction Resources
Mechanical Construction Resources
Mechanical Engineering Support
Electrical Engineering Support
Project Support
Industrial Engineering Support
Physical & Electronic Library
Administrative Support
Financial Support
Computer Support
Capital Planning
Operations Department A Projects
Operations Department B Projects
Operations Department C Projects
Business Unit A Projects
Business Unit B Projects
Business Unit C Projects
Customer Needs (Projects)

Figure 5.5

Old Organization Structure of Engineering Group

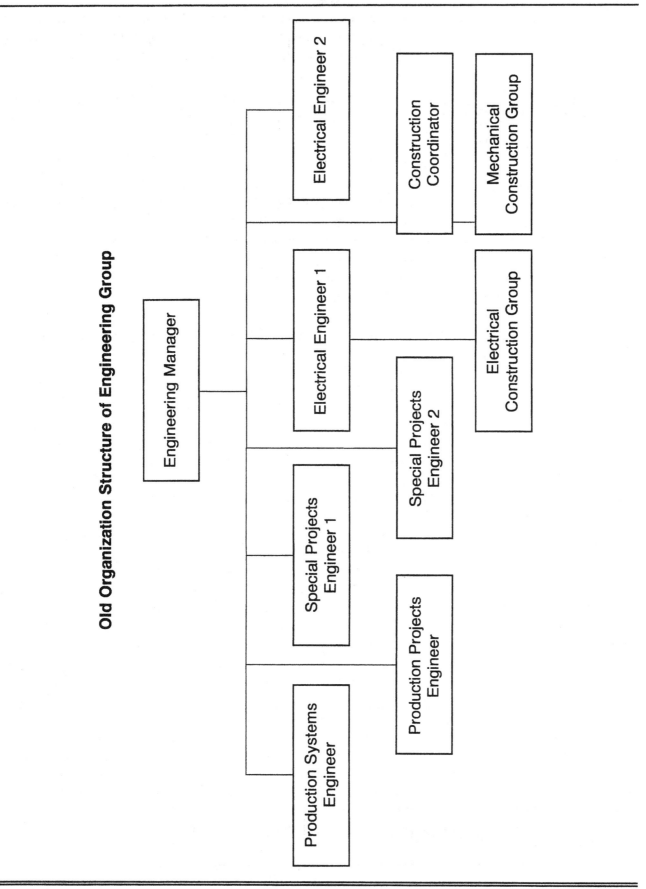

Figure 5.6

- Provides organization, accountability, and better communication. The project operations group, with project managers focused on particular functions, provides better communications.

The Project Life Cycle

Projects progress through a series of activities which can be grouped into major phases. These phases are known collectively as the *project life cycle*. An understanding of what happens in these phases enables the project manager to control the limited resources more effectively.

The level of effort, resources required, number of people (and their respective skills and abilities) and sources of conflict will all change as a project progresses through its life cycle. Generally, every project progresses through several basic phases which can be described in chronological order, as follows:

- Programming
- Preliminary (Schematic) Design
- Final Design
- Implementation (Installation, Construction or Production)

Old Organization Structure of Project Group

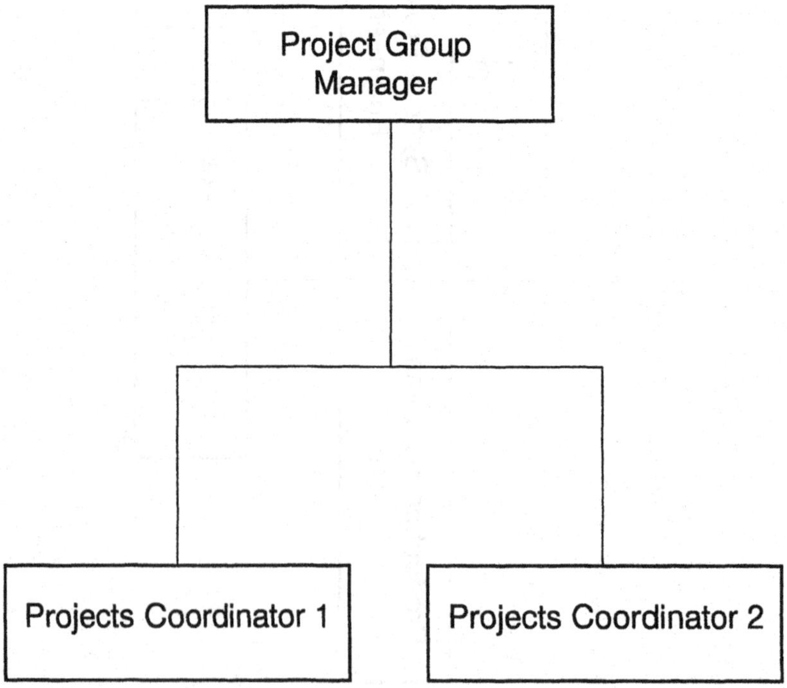

```
              ┌─────────────────┐
              │  Project Group  │
              │    Manager      │
              └────────┬────────┘
                       │
          ┌────────────┴────────────┐
          │                         │
┌───────────────────┐    ┌───────────────────┐
│ Projects          │    │ Projects          │
│ Coordinator 1     │    │ Coordinator 2     │
└───────────────────┘    └───────────────────┘
```

Figure 5.7

124

Using a Hybrid Organization Structure

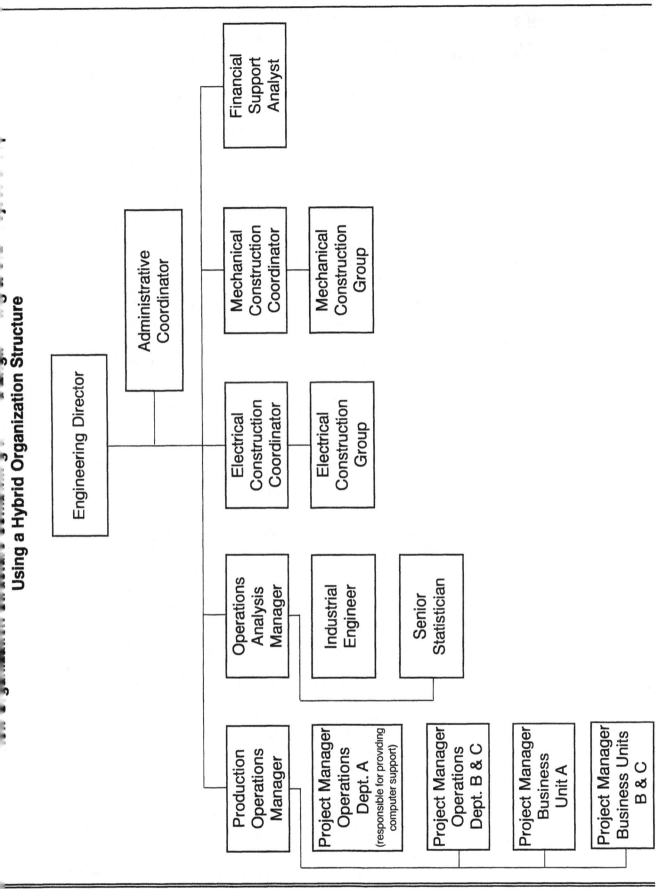

Figure 5.8

Role of the Project Manager

An essential factor in successful project management is a capable project manager with the authority and responsibility to do the job. Therefore it is necessary to employ, engage, or appoint an appropriate professional project manager. The success or failure of a project frequently depends on the project manager's ability to balance carefully a variety of technical, economic, social, and political influences. Time and again, the project manager selected to get a job done may be very competent in a particular technical area (such as engineering, finance, office administration, production, management information systems, maintenance, or property management), but lacks the expertise required for overall project management. Moreover, the person assuming responsibility for the project may be unable to devote the time necessary to handle the project details to completion.

The project manager's major role is managing multiple disciplines and limited resources to produce the desired product or service using the combined technical expertise of different functional disciplines. This manager must see to the completion of the project in conformance with requirements and specifications, on schedule and within budget. The project manager is ultimately responsible for ensuring that the expectations of the project customer are met or exceeded. The project manager is the official designated team leader for dealing with all in-house and external parties involved in accomplishing the project objectives.

Project Management Stress & Conflict

Project management involves getting things done through other people. It is no place for the timid. Projects by their very nature involve stress and conflict. The project manager has the responsibility to lead people over whom he or she often does not have direct authority to achieve the project objectives. Typically, the project team is comprised of friendly adversaries who are temporarily brought together to accomplish the project objectives. All view the project objectives from their own vantage points. In a construction project, the project team may be comprised of suppliers, architects, engineers, contractors, specialty consultants, project sponsors or customers, and members of the same company representing different departments or holding higher positions in the organization. The project manager must be sensitive to their needs and concerns and deal effectively with inevitable conflicts of interest in order to promote the interests of the project.

The successful resolution of conflict is one of the most challenging tasks a project manager faces, and one of his or her most essential skills. The project manager must continually be alert to potential conflicts in order to avoid or resolve them. Certain types of conflict are more prevalent during particular phases of the project life cycle.

In addition to potential personality conflicts, disagreements may occur regarding:

- Definition of the problem
- Program objectives and priorities
- Allocation of scarce resources (capital and manpower)
- Team member roles and responsibilities
- Technical opinions
- Procedures
- Reporting relationships
- Paperwork requirements
- Cost estimates
- Scope of work
- Scheduling

Conflicts in the early phases of a project life cycle tend to revolve around the program objectives, procedures, cost estimates and responsibilities, while conflicts regarding schedule and scope of work tend to surface in the later phases of the project life cycle.

Much of the project manager's success hinges on his or her ability to build an effective team. An effective project manager has a systematic problem-solving approach. He or she must uncover and confront conflicts and inappropriate behavior promptly to avoid undermining the team effort. The project manager needs the creativity and ideas of all team members. Since fear undermines productivity, preventing people from being innovative and sharing ideas needed to arrive at solutions, the project manager must try to control it. When there is a problem, the effective project manager is careful not to overreact emotionally and seek someone to blame. He or she must be cognizant that most problems within an organization are due to processes rather than people. It is therefore necessary to identify and solve the problems that will invariably occur first by reviewing the process.

The effective project manager works well with other people, and exercises leadership, sales and negotiation, communication, consensus-building, problem-solving, and organization skills. These must be combined with a sense of urgency about getting things done in a timely fashion to accomplish the project objectives.

Project management is a team concept. Every leader knows that power is critical to getting people to work together as a team. The more power a leader has at his or her disposal, the more likely they are to get the desired results. Project managers must be able to identify and capture the sources of available power to get other team members committed to support the project. Some sources of power that the project manager can use to influence team members include:

1. *Formal authority* or high position in the organization which gives the person the ability to select team members, make payments, and reward and punish.

2. *Technical expertise* or information which enables him or her to influence other team members.

3. *Influence* (close relationships with customers, suppliers, or other people inside or outside of the organization): Relationships with more powerful people enables the project manager to influence the career path of team members.

The best project managers use their own leadership ability, interpersonal skills, and charisma to enhance their authority and obtain the active support of the people needed to achieve the project objectives.

How to Plan and Scope Projects

Failure to plan projects carefully can result in excessive rework, unnecessary costs, schedule delays, quality deficiencies, disputes, and even litigation. Many projects fail because the project manager or owner prematurely makes a big decision requiring a "leap of faith." He or she commits full funds or awards contracts for a project when the project requirements have only been partially defined. There remain too many unknowns to quantify and control the risks. Because of time constraints, project stakeholders may be pressured to make important decisions with very little information. (Stakeholders are individuals or entities who could affect or are affected by the project.) How, under these constraints, does one optimize control of time, cost and quality in delivering a project?

The difficulty in planning corporate investments in facilities is that you can never figure out exactly what the future holds. To increase the probability of project success in an uncertain world, you must treat the big decisions like playing stud poker. With each card you invest some money and you get more information. At the end of each phase of a facility project, you get more information—like another card. Towards the third or fourth "card" you either get close to finalizing your planning decision, or you decide to fold.

The completion of each phase may be thought of as a "milestone" or "turning point," at which time the project manager presents "deliverables" to the project customer. "Deliverables" are the results of project assignments that lead up to a milestone.

At each milestone, the customer is presented with information he or she needs to make informed decisions. For example, at the first milestone—completion of project planning and programming—the project parameters are defined in terms of financial, technical, quality and time. The milestone approach offers the customer the benefit of progressively releasing funds for the project as he or she obtains more information and greater assurance that the objectives can and will be achieved. It provides the opportunity to redirect the project or cancel further involvement before the start of another phase.

Focusing on Customer Requirements

Projects are undertaken to solve problems or satisfy needs. Many projects have failed because no systematic inquiry of end users' requirements was conducted. Or, the project selected was not the solution to the root cause of the customer's problem. Typically, customers want more than they can afford; it is therefore important that the project manager distinguish needs from wants, and devise a strategy that provides the best values. The project manager's primary responsibility is ensuring that the end product meets the client's requirements. The project manager must clarify, document, and prioritize the functions and objectives of the project. Once the problem or need is stated, the project objectives can be defined.

During the project planning and programming phase, the needs of the customer or user groups within the facility, and the scope of work are defined. The data gathered during this phase becomes an important part of the project record. The focus should be on cost, time, quality, and technical requirements. Technical requirements may include space, utility, equipment, climate control, and environmental requirements for each department. A key ingredient to good project management is obtaining the necessary information to prepare a realistic project plan.

Value engineering is a powerful method for determining the requirements for a project and separating customer needs from wants. Function Analysis System Technique (FAST diagramming, as described in Chapter 4) is a practical tool for getting the project team members to buy into what the program is all about during the planning phase of the project.

Defining the Project Scope

Studies have shown that poorly defined customer/user requirements and allowing work to exceed the project scope (often referred to as "scope creep"), are two factors that contribute the most to project cost overruns. The scope of work can be used to:

1. Communicate the project's purpose and objectives.
2. Provide a thorough description of the project including: parameters, activities, processes, systems, equipment, and support requirements.
3. Lead to approval of essentials that fix and describe the project's size and character, before proceeding with design development.
4. Illustrate possible solutions within the shortest time frame and at minimum expense.
5. Determine the feasibility of proposed project ideas.
6. Establish relationships among various systems and components on a first-cost and life cycle cost basis.
7. Prepare a project estimate and schedule.

Scope management is the function of controlling a project from concept to completion. The first step in effective scope management is defining and documenting the project objectives and selecting the best approach to achieving them. It is necessary that the project manager develop a written project scope statement. This statement is a documented description of the justification for the project, the work content or components of the project, the major deliverables, and the project objectives. The project manager should refrain from preparing the project scope statement until he or she understands the customer/user requirements and project objectives.

Following is an example project statement:

A. This renovation project will bring into compliance all current violations of the Americans with Disabilities Act of 1990 as identified in an ADA audit of the facility dated September 1, 1992. This audit was performed by an interior design company under a separate contract.

A major feature of the project is to provide a passenger elevator within the facility to access the basement, first, and second floors.

B. Constraints

The existing office must maintain full operation during construction.

If the building structure is to be opened, measures must be taken to keep out rain, snow, and wind. The construction must be scheduled to avoid work with the structure open in the winter.

Defining Project Objectives A project objective is a statement specifying the end results desired. These statements form the foundation of the planning process and should be reviewed frequently. Objectives should be monitored and updated continuously to reflect the best understanding of the client's needs.

During the design phase, objectives are translated into specifications for each component of the project. Objectives define the project's specific deliverables and expected outcomes. Sometimes, as a project progresses, objectives may have to be revised. Failure to update the objectives and communicate the changes to project team members will result in an imprecise definition of the project scope and poor scope management. It will also lead to failure to satisfy the customer' needs.

Work Breakdown Structure

Once the problem is stated and the customer/user requirements and project objectives defined, a work breakdown structure can be developed. The work breakdown structure is a tool the project manager uses to manage the project. It defines, organizes, and communicates the planned scope of work to achieve the project objectives. The work breakdown structure is simply the list of activities (tasks) necessary to accomplish the project objectives. These activities are organized in a logical manner. A work breakdown structure may be thought of as a to-do list of project activities to be performed. Figure 5.9 is a sample request for proposal for engineering analysis identifying the justification for the project, background, objectives, and work breakdown structure.

The project manager must select a point of view from which to define and organize the elements of the work breakdown structure. The following list is an example of how these elements might be organized

- Parts or components
- Functions
- Organizational disciplines or departments
- Responsibilities of project personnel
- Location
- System
- By established standards (such as classifying construction costs by C.S.I. MasterFormat)

The planner's preference or generally accepted practices determine how the work breakdown structure is organized.

The project manager breaks down each project phase and sub-phase into activities until a sufficient level of detail is reached to manage the activities effectively. Many project managers consider the work breakdown structure the single most important project document. It is the basis for estimating costs and developing budgets, scheduling activities, organizing project resources, and assigning responsibilities to team members. It also provides a framework for decision-making and developing the project control system. Preparation of the work breakdown structure helps to identify project risks. It enables the project manager to develop a plan and a strategy to address the risks, thereby increasing the probability of success.

The people who will be performing the activities should be involved in planning the work. The planning group should confer in a brainstorming session, arriving at a work breakdown structure that reflects the project team's concept of the project as a whole. Elements in the work breakdown structure must be broken down to a level relevant to the people who will be using it. For example, customers may need only a general summary of project activities, while the project staff will need a more detailed list of the activities planned to complete the project.

Communicating the Scope of Work

It is the project manager's responsibility to communicate the scope of work to the people who will carry out the activities. The completed project scope should fully describe the work to be performed, the resources consumed, and the results. It is important to define quality standards in the project planning documents.

The scope of work is also used to estimate the cost to complete a project. Changes or ambiguity in the scope of work can increase the

Proposal to Conduct an Engineering Analysis of
Cause of Product Defects at Plastic Molding Line Ovens
(For a Manufacturer of Plastic Containers)

Conduct an engineering study to determine if the product quality problem at the plastic molding ovens in the northwest corner of the Plastics Room is due to a draft condition, and to provide options for addressing the draft condition if it is found to be the cause.

BACKGROUND

During our meeting, we discussed the operation of the heating and ventilating equipment in the space, the usage of the shipping and receiving dock and the resulting conditions at the plastic molding oven. The production line foreman identified a persistent product quality problem during the winter at two ovens near a door to a warehouse area. This warehouse area adjoins the shipping and receiving dock, and the line foreman believes that cold drafts through the mold room door are causing the problems at the plastic ovens. Review the situation and offer recommendations. Any solutions should address the cause of the problem, not just the symptoms.

OBJECTIVE

1. Determine whether the cause of the product defects is a draft, as proposed by the production line foreman.
2. Identify cost effective conceptual solutions for eliminating the draft and/or its impact on the manufacturing process.

NOTE: During the early stages of the study, if findings indicate that the product defects are not related to a draft condition, notify us immediately and await your instructions before proceeding any further.

WORK BREAKDOWN STRUCTURE

1. Meet with the production foreman in charge of operating the particular process line in order to precisely determine what the product quality problems are, when they occur, and what the observable space conditions are at that time.
2. Review the equipment operating conditions and observed problems to help verify the cause of the product defect problem.
3. Meet with Building Services personnel to review the operation of the heating, ventilating, and process exhaust equipment in the space. (Note that Items 3 through 6 are based on the assumption that the product defects are caused by a draft problem. If the initial review indicates otherwise, the engineering firm will notify the Owner immediately and will stop work, pending further instructions.)
4. Review drawings and heating and ventilating equipment information (provided by Owner) to determine building pressurization in the molding room and the adjoining warehouse in order to identify the cause(s) of the draft at the process ovens.
5. Evaluate various retrofit alternatives for eliminating the draft (or its impact on the production process).
6. Prepare a summary report identifying its cause and the most cost effective remedies.

Figure 5.9

cost and delay the schedule. Clearly, the scope of work should be well defined and realistic.

The scope of work is the first issue that should be addressed in contract negotiations, yet, too often, it is not given proper attention. Many disputes can be avoided by proper preparation and documentation of the project scope of work.

"Control of Construction Project Scope, A Report for the Construction Industry Institute" by O'Connor and Vickroy[1] states that owners are sometimes willing to proceed with a project without adequate scope definition because:

- They lack sufficient engineering expertise to provide a complete conceptual definition;
- They believe there is no reason to spend additional funds in scope definition if the initial Order of Magnitude (feasibility) estimate looks favorable;
- As a matter of economy, they want to limit the amount spent on feasibility and project authorization and budget studies;
- They think that the shorter the time spent on preliminaries, the quicker the job can be completed, and the lower the overhead;
- High interest rates and market pressures make project duration critical.

Documentation

Documents identifying the project scope of work should be included as exhibits in contracts between owners and vendors, contractors and consultants. The scope and nature of the services required can be communicated in the form of either plans and specifications or a statement of performance requirements.

Scope documents for a construction project should identify known information, sources and assumptions, such as:

- Background
- Project objectives
- Site description
- Site development requirements
- Facility layout and functional relationships
- Flexibility requirements
- Current and future requirements for equipment and space
- Functions and priority requirements
- Vertical and horizontal space requirements for each department
- List of equipment, including dimensions, weight, location, heat rejection, exhaust, and utility support requirements (such as compressed air, chilled water, gas, etc.)
- Electric power and lighting requirements
- Requirements for each department (e.g., special flooring, foundations, noise control)
- Number of occupants and occupant activities

- Environmental control scheme and desired space conditions, including temperature, humidity, air filtration, air pressurization, exhaust, and zoning requirements for each department
- Plumbing and sewage requirements
- Fire protection, detection, and life safety requirements
- Security requirements
- Milestone schedule
- Budgetary description and requirements

Detailed data is needed on each piece of equipment and support requirements. For example, motor horsepower, nameplate amperage, and operating amps, volts, and diversity factor (the actual hours of operation as a percentage of time) are needed to calculate heat gains and losses for designing air conditioning and heating systems and to determine electrical requirements.

Figures 5.10a and b are examples of equipment inventory sheets that can be used to record information for this purpose. Figure 5.10c is a facility programming worksheet for offices and support areas.

Manufacturers' data (including equipment dimensions and weights, space, structural, power, utility, energy, and environmental requirements) is needed. Design of a plant's heating, ventilating and air conditioning (HVAC), plumbing, electrical, and structural systems depends on many factors. These include locations, dimensions, and weights of equipment; requirements for power, temperature, humidity and zoning, and ventilation; and exhaust and heat load information. Process requirements for chilled water, gas and compressed air also should be considered.

Equipment data can be obtained from manufacturers' specifications, interviews with manufacturers' representatives, or by visually observing the equipment. Be sure to provide ample space for operator activities, regular cleaning, maintenance, and service procedures, as well as for support requirements such as ductwork, utilities, electrical and sprinklers.

The objective is to approve the scope and essentials before proceeding with more elaborate drawings and specifications, which fix the project's size and character. Scope documents may consist of single-line schematic drawings and outline specifications for the building systems, each of which may include the following items:
- Site plan
- Floor plan
- Building elevations and cross-sections
- A written description of the project program requirements including the basic building systems, financial, and schedule requirements

Figure 5.11 is an outline of items to be included in a request for authorization for a project. Figure 5.12 is an outline for preparing a project work plan.

Summary A project-driven facilities department can meet customer needs, despite the challenges of smaller staffs and declining operating budgets. Selecting the appropriate organizational structure depends on the problems that need to be addressed. Involving project team members in defining and documenting the project requirements (scope of work) is a prerequisite to efficient use of scarce resources and obtaining team commitment to project and organizational objectives.

1. "Control of Construction Project Scope, A Report for the Construction Industry Institute," O'Connor and Vickroy, University of Texas at Austin, 1986.

The Sievert Corporation
Founded in 1917

Engineering/Construction/Project Management/Facilities Planning
A Member of the Sievert Group
2095 Hammond Drive, Schaumburg, Illinois 60173

FACILITY PROGRAMMING WORKSHEET
Machinery and Equipment Data

Manufacturer: _____ Model Name & No _____ Machine Tag Number. _____

Prepared By _____ Date _____ Department _____

Machine Function. _____

Adjacency Requirements: _____

Physical Requirements:

Length _____ Width _____ Height _____ Weight _____

Clearances: Front· _____ Back· _____ Left _____ Right: _____ Above. _____

Electrical
Nameplate Data (Volts, Amps, Phase):
_____
_____
_____

Natural Gas: Yes (_____) No (_____)
CFH: _____ Pressure. _____

Compressor Gas: Yes (_____) No (_____)
CFM _____
PSI _____

Vacuum: Yes (_____) No (_____)
CFM. _____
IN.Hg _____

Chilled Water: Yes (_____) No (_____)
Inlet Temp. Range. _____
GPM. _____
Temp Rise: _____
Pressure Drop _____
BTU/Hr _____

Condensate: Yes (_____) No (_____)
PSI. _____
Returned. _____
Wasted: _____

Remarks: _____
_____
_____
_____

Heat Rejection
BTU/Hr _____
Oper. Amps (Running/Idle): _____ / _____
Under Hood (Yes/No): _____
Mach. Run Time (Max. Hrs/Day) _____
Mach. Run Time (Max. Min./Hrs): _____
Mach Run Time (Max Contin. Operations):
_____

Exhaust: Yes (_____) No (_____)
CFH _____
Press Diff (IN W C): _____
Required Treatment _____

Vent/Stack: Yes (_____) No (_____)
CFH _____
Temperature· _____
Required Treatment· _____

Domestic Cold Water: Yes (_____) No (_____)
GPM _____
PSI _____

Drain: Yes (_____) No (_____)
GPM. _____
Required Treatment: _____

Figure 5.10a

Print Center Equipment Specifications

Machine #	Quantity	Room #	Title	Nameplate		Operating			Heat Dissipation (BTU/HR)
				Volts (V)	Amps (A)	Volts (V)	Amps (A)	Phase	
1	2	130	IBM 3170	240	50	210	22.2	3	27,454
1a	2	130/131	Cooling Unit	240	20	209	8.5	1	5,738
1b	2	130	Console	120	20			1	1,000
2	2	130	Xerox Docutech 135	240	50	209	40.3	1	25,773
			Scanner	240	30	209	15	1	9,593
			Server	120	20			1	1,000
3	1	130	Heidelberg	208	50	**	**	3	21,420

The Sievert Corporation
Founded in 1917

Engineering/Construction/Project Management/Facilities Planning
A Member of the Sievert Group
2095 Hammond Drive, Schaumburg, Illinois 60173

FACILITY PROGRAMMING WORKSHEET
Offices and Support Areas

Project _____ Project No _____

Client· _____ Prepared By· _____ Date· _____

General
Name/No _____
Dept. _____
Access/Conn _____
Sq Ft _____

Usage
Times (8-5, 24/day, etc.) _____
Days (M-F, 7/wk, etc) _____

People
Qty. (Typ./Max.) _____
Activity _____

Environment
Summer (Temp./rh) _____
Winter (Temp /rh) _____

Lighting
General
  Illum (Fc or W/S.F.) _____
  Type _____
  Switching _____
Accent
  Spacing _____
  Type _____
  Switching _____
Other (Describe) _____
  Spacing _____
  Type _____
  Switching _____

Communications
Telephone _____
Fax _____
Data _____

Equipment
Computer (Qty.) _____
Monitor (Qty.) _____
Printer (Qty) _____
  Oper Amps _____
  Idle Amps _____
  Max. Oper (Min./Hr.) _____
Fax (Qty) _____
  Oper. Amps _____
  Idle Amps _____
  Max. Oper (Min./Hr) _____
Other (Qty and Descr.) _____
  Oper Amps _____
  Idle Amps _____
  Max. Oper (Min./Hr.) _____
Other (Qty. and Descr) _____
  Oper. Amps _____
  Idle Amps _____
  Max Oper (Min./Hr.) _____

Electrical Outlets
120V/1Ph (Qty) _____
208V/1Ph (Qty) _____
208V/3Ph (Qty) _____
220V/1Ph (Qty) _____
___ V/ ____ Ph (Qty) _____

Furniture/Fixtures
Descr /Dim /Qty _____
Descr./Dim./Qty _____
Descr /Dim./Qty _____
Descr /Dim /Qty. _____
Descr /Dim /Qty. _____

Comments: _____
_____
_____
_____
_____

Figure 5.10c

Outline for Project Proposal

1. Project title/date
2. Project team and manager
3. Purpose of this project: strategic objective being supported
4. Project objectives
   4.1 Expected results
   4.2 Sequential list of major milestones
5. Short paragraph on project description
6. General approach on underlying description
7. Benefits/advantages
8. Constraints/limitations
9. How long is it going to take to do this project?
10. What type of resources will be needed?
    10.1 People skills
    10.2 Special equipment
    10.3 Special facilities
11. Estimated project cost
12. Team qualifications

gure 5.11

Work Plan Outline

1. Project identification
   Project name
   Project team
   Date
2. Scope of project
   Project mission
   Objectives
     Financial
     Products and services
   Work breakdown structure
3. Quality of work
   Statement of general quality of work
   Specific quality criteria/specifications
4. Project schedule
   Project Gantt chart (software generated)
   Project network diagram (software generated)
   Activity status report
5. Environmental analysis
   Internal environment
   External environment
6. Organizational structure and style
7. Project team responsibility matrix
8. Flow charts
9. Communications and information
   Scheduled meetings
   Emergency meetings
   Project management information systems to be used
   Report shedule and distribution matrix
10. Project cost
    Task budgets
    Total project budget
11. Risk analysis
    Risk to company if the project fails
    Risks which could damage or destroy the project
    Criteria for practical recovery from crisis
12. Contracting/procurement
    Obtaining internal services and personnel
    Obtaining external services and personnel

gure 5.12

Chapter 6

Cost Management

A primary function of facility management is to control total building life cycle costs and to keep them within budget constraints. Some basic cost management tools help facility managers ensure that facilities are managed in the most economical manner, consistent with the program requirements, performance benchmarks, budget and schedule restraints. These tools include: cost estimating, value engineering, work breakdown structures, life cycle costing, design-to-cost controls, contracts, schedules, and cost monitoring and reporting systems.

The major facility life cycle cost categories are acquisition, design, construction, operation, maintenance, and disposal. Other costs include, but are not limited to, rent or mortgage, property insurance, real estate taxes, furnishings, equipment, code compliance, utility, security, waste disposal, and custodial costs. The largest facility operating costs typically include rent or mortgage payments, depreciation, utilities, and real estate taxes.

Some facility costs vary directly with sales and production rates, such as power consumption, waste disposal, and production equipment maintenance costs, while other costs continue with a much looser relationship to building activity (light, heat, cooling and ventilation, building maintenance, and janitorial services). Management must evaluate facility costs for conformance with established standards and cost management objectives.

Facility managers must develop realistic budgets and maintain control of capital and facility-related operating costs. Thoughtful and meticulous preparation is needed before making final decisions on projects. Money should be appropriated for feasibility studies, scope development, and budget estimates that properly reflect the project scope of work before committing full funds to a project and before major contracts are signed.

Cost management must be applied during every phase of the project life cycle. Many project failures are due to poor cost management during the early project phases when there is the greatest ability to control the overall costs of a project. This is the time when the scope rather than the details of the project are defined. The ability to influence cost savings diminishes with time — from 100% during the early planning phase to perhaps 5% at the start of construction. In addition, the cost of making changes (even those designed to save money) increases as the project progresses. Once the construction c installation phase begins, contractors are merely following the plans and specifications. Contracts have been let, material and equipment have been ordered, and labor hours have been committed. Figure 6.1 is a graph that shows the diminishing ability to achieve savings as the project progresses through its phases.

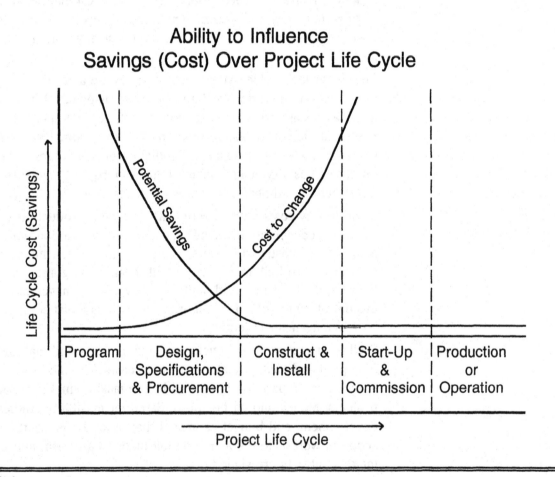

Figure 6.1

Estimating Project Costs

Cost estimating is the basis of cost management. Cost estimating is the systematic process of assembling, evaluating, and forecasting the costs of a project. The completed cost estimate is based on the approved scope of work and project objectives. The cost estimate serves as a target cost. It is used as a benchmark for cost control during the life cycle of the project. Estimating is a statistical art rather than an exact science. The following guidelines can help you do a better job of estimating.

Guideline #1

Before preparing an estimate, it is important to have a clear understanding of the project objectives and each element of the work breakdown structure. The estimator must thoroughly evaluate factors that are likely to influence cost (personnel resources, time, methods, materials, location, equipment, facilities, contractors, local economic conditions). This is done by having a detailed and complete scope of work as defined by the Work Breakdown Structure, design documents, and specifications. If there are errors or omissions in the Work Breakdown Structure, conceptual design or specifications, the project economic feasibility study will be flawed. If the scope of work has not been defined sufficiently, estimates will be too low and the project will exceed budget expectations. Consider the various types of costs — fixed, variable, direct and indirect. Other costs to be considered in any cost analysis are sunk and opportunity costs.

- **Fixed costs** do not change with variations in production output or volume. Examples are overhead costs such as property insurance, real estate taxes, landscape maintenance, housekeeping, and salaries of department managers.
- **Variable costs** tend to change with output. Direct material, direct labor, and operating supplies are examples of variable costs.
- **Direct costs** are assignable or can be traced to a specific process, product, or project. Examples are materials or equipment that become part of an installation. Operating and project personnel assigned to a project or service may be considered as direct costs.
- **Indirect costs** are not directly assignable or cannot be clearly traced to the end product, project, process, or service. Indirect costs may include certain types of insurance, property taxes, mortgages and lease costs, and maintenance.
- **Sunk costs** are costs which are committed and cannot be recovered.
- **Opportunity costs** represent the alternative investment opportunities that are foregone because available funds have been allocated to something else.

Guideline #2

Cost estimates have a cost of their own. The time, effort, and cost of preparing an estimate increases significantly with the accuracy required. As the scope of work and design develops, unknowns are identified and cost estimating accuracy increases proportionately. The problem is that many project owners need to know how much th total project will cost before the design is complete. Methods to avoid this problem will be discussed in Chapter 8, "Contracting and Procurement Methods." A project schedule is also necessary to establish an accurate cost estimate, as many costs are time-related. Figure 6.2 is a graph showing the decrease in undefined project cost a the project progresses and there are fewer unknowns.

It is not uncommon for owners to establish project budgets by soliciting free input from design consultants or contractors. Often thi is done without the benefit of detailed and complete documents reflecting the project scope and quality of work. This practice is risk because there is a lack of firm or verifiable information for the evaluation of project costs.

Designers or contractors may provide a "cost estimate" stated in terms of dollars per square foot of building area based on their experience and judgment, historical cost data from similar projects, c rule-of-thumb calculations. Because every project is unique and ther are many variables that can affect the cost of a facility, estimates

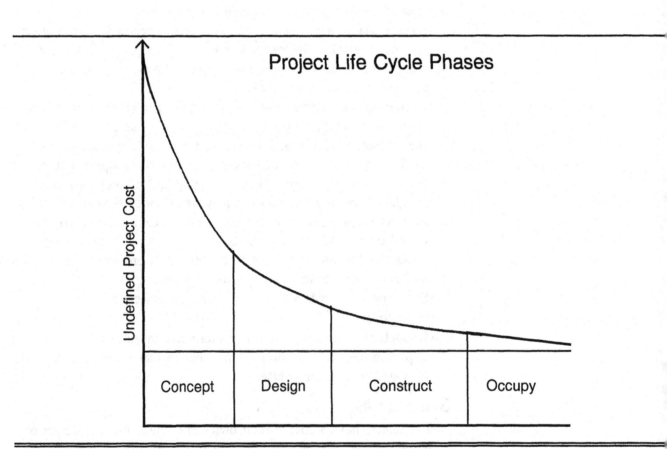

Figure 6.2

144

based on square foot of building area or volume alone may result in oversimplification or generalization. The accuracy range of conceptual, "ballpark," or "order of magnitude" type estimates based on square feet of building area or similar past projects tends to be very low. This type of estimate should be used only as a very quick and low-cost way to screen potential projects and to decide whether a more accurate, detailed quantity survey of each building system and component is justified.

Historical data alone should not be employed for estimating project costs. Many factors can differ from one project to another, including:

- Location and accessibility
- Quality expectations
- Scope of the project
- Competitive and economic conditions
- Soil and subsurface conditions
- Contractual requirements
- Time of project and weather conditions
- size of the project
- Productivity rates (which may be affected by learning curve and crew composition)

Similar past projects may serve as rough guides at best. The estimate, like the project work breakdown structure, must be broken down into successively finer detail to increase estimating accuracy. Without an accurate and complete scope of work, consultants and contractors may be inclined to be overly optimistic. The resource requirements may be understated, and project costs underestimated. This situation sometimes results from a fear that the project could be cut if the budget estimate is set high. It is a mistake to obtain quotations without giving outside consultants or contractors detailed, accurate, and complete scope documents. In some cases, consultants and contractors may submit quotes on the low side to stay under consideration. Without supporting documentation, cost estimates will be questionable. Beware of firms that have a reputation for giving low bids to get work and then fail to deliver. It should be understood that a consultant or contractor must make money or go broke.

Guideline #3

To verify that the project will remain within budget, it is recommended that estimates be updated as the design progresses. Preliminary cost or conceptual estimates are prepared during the feasibility and early planning phases of the project. Once a well-defined scope of work has been achieved and the design is about one-third complete, an improved cost estimate can be developed. As the design develops, more detailed and accurate estimates can be made to guide further commitment of funds. Updating and refining the estimate as more information becomes available serves as a reality check and helps to avoid "budget creep" during the design phase. If the cost estimates

145

prepared during the various phases of the design process identify that the project scope has grown larger than the project owner's budget, the project team members must generate ideas for reducing the cost back to the owner's planned budget. The planning and design process for a construction project can be broken down into the following phases:

- **Programming** – The first phase of the project life cycle which defines the performance requirements for the project design — including technical, financial, and time requirements. The design criteria are defined in both written and graphic terms.

- **Schematic Design** – The process of translating the program performance requirements into simple concept drawings illustrating the general scope of work for construction or installation of the primary building systems. The schematics are used to prepare a preliminary construction cost estimate which is sometimes called a *schematic* or *conceptual estimate*. Single-line sketches are commonly used to describe the floor plan concept, site plan concept, exterior elevations, and the building engineering systems.

- **Design Development** – The process of developing and refining the schematic drawings approved by the owner for the various building systems. These documents fix and describe the architectural, structural, mechanical, electrical, and process requirements of the project. When the design development documents and estimate are approved by the project owner, they become the basis for construction documents.

- **Construction Documents** – The complete set of construction documents communicates in detail the requirements for the entire construction project. The construction drawings and specifications are suitable for preparation of detailed estimates, permit procurement, obtaining competitive bids from contractors and for performing construction.

The cost estimate becomes the source for development of a budget and the basis for requesting appropriation of funds. The budget is a plan for allocation of available funds. It is the standard that is used to determine if the project manager did a good job with managing the company's financial resources in completing the project.

Project Budget Checklist

The project manager can help develop a realistic budget based on the scope of work and feasibility studies. A budget checklist should be designed and become part of the project record. While broad in scope, the following checklist could be applicable to your project budget:

1. Site acquisition costs
2. Demolition costs

3. Architectural, engineering, project and construction manager, and other professional service fees
4. Owner's in-house staff costs
5. Site work
6. Utilities
7. Construction
8. Construction testing service fees
9. Survey fees
10. Financing
11. Legal fees
12. Taxes
13. Equipment
14. Furnishings
15. Insurance and bonds
16. Landscaping
17. Security systems
18. Communications systems
19. Overtime work or delaying the project
20. Moving and transportation costs
21. Permits
22. Parking lot paving and striping
23. Cleanup
24. Contingency allowance

Cost Information Sources

The MasterFormat developed by the Construction Specifications Institute is an organized checklist of construction categories and a useful standard format for classifying data. It describes the systems and components that comprise a construction project. It is also a useful checklist for preparation and organization of specifications and cost estimates.

Usually budget estimates are prepared by applying unit costs (e.g., per square foot, cubic yard, lineal foot) for each building system and multiplying them to the quantities required. These costs can be obtained from the cost files of similar projects, from contractors, or from estimating guides such as *Means Building Construction Cost Data* or *Facilities Construction Cost Data*. Published statistical averages should be carefully used only by those who are knowledgeable in construction. This data can be adjusted for location, quantities, productivity, complexity, time frame, weather, and market conditions. Appendix B of this book contains an example of a conceptual construction cost estimate broken into the 16 CSI MasterFormat divisions.

Any estimates developed during the conceptual stage, when the scope of work is still general, should be subject to a large contingency factor. A contingency is an allowance intended to cover the cost for as-yet unknown items that would be under the project's defined scope of work covered by the estimate. These items could not have been defined at the time of the estimate because of lack of complete information. The contingency allowance is established separately as component in the estimate and carried as a line item percentage. Contingencies could cover such items as testing services, overtime work, unforeseen subsurface conditions, or concealed building conditions.

Monitoring and Reporting Costs

Once the project has been divided into the work packages, a code or numbering system is assigned to the activities for cost monitoring, control, and forecasting purposes. The accounts code allows monitoring of actual costs vs. budget — which is one of the main responsibilities of cost management. This process is carried out continually to identify problems and take corrective measures. It helps prevent small problems from turning into big problems. Variance analysis is used to identify problems, and reasons for the variance are determined so corrective measures can be taken.

Objectivity and a well-defined project scope are essential in order to remain in control of costs. It is necessary to avoid understating requirements in an effort to sell or justify the project during its conceptual phase.

Value Engineering

Value engineering, used during the conceptual phase of a project is a powerful cost management tool that demonstrates a cost avoidance effort. Value engineering cost avoidance can be applied to minimize direct and indirect costs and improve delivery of design and construction projects.

On major projects, you may want to conduct a formal value engineering study to determine the project's essential elements and how to achieve them most cost-effectively. The study should follow value engineering job plan phases: information-gathering, creativity, evaluation, development, presentation, and implementation. Studies are often conducted during the preliminary design phase before the major systems are selected and locked in. If the study is delayed and the design progresses further, substantial redesign may be necessary to incorporate identified cost-saving changes. These changes are more expensive to make as the project progresses. Most often, the cost of the value study is recovered by the resultant savings and efficiencies.

Cost Trade-Off Studies

Prior to design development, sketches of alternative conceptual design schemes should be prepared. Cost estimates should be prepared for each alternative to ensure the best selection of systems and components. Once systems are selected, design professionals can develop detailed construction drawings from which contractors can prepare detailed estimates and submit bids. These studies optimize value in the design stages when the opportunities to control costs are greatest.

Design professionals should develop designs in accordance with the budgeted dollar amount. According to the "design to cost" concept, cost is a parameter in the same sense as other technical performance parameters. Designs must be developed within the budgets for each line item. This will help reduce the need for redesigning major portions of a project and subsequent project disputes and delays.

The more detailed and precise the construction documents, the more accurate the contractors' bids. The result: fewer costly change orders in the field during construction and tighter project cost and schedule control.

Building Economics

Facility management involves trade-offs, such as life cycle versus initial costs. While businesses are often looking for top quality, many do not want to budget for it. However, we know that looking at first costs only can lead to inferior performance, higher maintenance and operating costs, and possible disputes. Over the life of the facility, operating costs generally exceed the original capital investment. Therefore, these expenses should be analyzed during the planning and early design phases of the project.

Very few projects are embraced by management and automatically funded. With the exception of legal mandates (e.g., EPA, OSHA, or ADA compliance), simply saying that the initiative has merit and is good for the business will not produce the necessary financial and personnel resources necessary for implementation. Your project will have to prove itself worthy of corporate support and it will have to be sold to management in the financial terms that management understands best.

Your proposal will be competing against numerous dissimilar initiatives that produce savings and/or revenues. For example, your energy conservation program may be competing against a proposal for more efficient production equipment or the introduction of a totally new product line. To enable management to choose between these dissimilar proposals, each must be expressed in common terms, the terminology of money.

Financial evaluators are not just for upper management's use. Like corporate management, you too must have a set of evaluators that can identify those projects that will be of greatest benefit in meeting your department's goals. Since you are either aiming to maximize profits or minimize the expenses required to produce your services, you too need financial/economic tools to aid in decision-making, even when you are in control of the money and other resources.

Financial Evaluation Techniques

The evaluation techniques range from simplistic to complex. Each has its uses and limits. For example, "Simple Payback" requires relatively little data and can be accomplished quickly. Its principal weakness is that it fails to account for the time value of money and the timing of cash flows — potentially critical factors. At the other extreme is the "Life Cycle Cost Analysis" which includes many additional factors and requires more time and effort to perform. The method you use will depend on the audience to whom you will be making the proposal, and the terms that are meaningful to them.

During the early phases of the project analysis, you would likely use the more simplistic methods to determine if a proposal has sufficient merit to justify additional study or development effort. You can use "Simple Payback" to determine which, if any, of several alternatives should be pursued. "Simple Payback" (sometimes called "undiscounted payback") is a calculation of the time it takes for a project to produce sufficient net income (or savings) to recover its initial investment cost. Its calculation requires knowledge of the project's first costs and the resulting expense reduction and/or net income streams.

Simple Payback = Investment/Annual Net Income

or

Simple Payback = Investment/Annual Net Savings

For example, if you invested $10,000 in a project that reduced your expenses by $2,000 per year, it would take five years to recover the initial investment ($10,000 Investment/$2,000 Annual Net Savings = 5 Years). This is also referred to as "the point when the cash flow turns positive," i.e., the point when more money has come into the business than has gone out as a result of the project.

This process becomes a little more complex when the net income or net savings are not constant from year to year. However, using a time line or cash flow chart, it becomes relatively simple to determine the break-even point. Suppose you made a $10,000 investment that had the following income characteristics.

Year	1	2	3	4	5	6
Net Income/Year	2,000	3,000	5,000	7,000	8,000	9,000

The simple payback is determined by finding the year when the aggregate of the income equals the initial investment.

Year	1	2	3	4	5	6
Net Income/Year	2,000	3,000	5,000	7,000	8,000	9,000
Aggregate Net income	2,000	5,000	*10,000*	17,000	25,000	34,000

Each company and each department will have to establish its own "simple payback" criteria for rejecting or accepting a proposal. The criteria could be absolute or variable, depending on circumstances. Any project that exceeds the maximum payback (say five years) would be automatically rejected. However, projects that have shorter paybacks may also be rejected if there are too many to fund from the current budget. When there is an abundance of good projects, the maximum payback cut-off date may drop to four, three, or two years!

A more complete financial analysis of investment alternatives considers life cycle costs and the time value of money. Life cycle cost analysis includes all costs, present and future. The alternative with the lowest life cycle cost will make the biggest contribution to the bottom line.

Life Cycle Cost Analysis

Basing decisions solely on lowest initial or first costs frequently leads to the implementation of plans that are not the most economical. For instance, a low efficiency motor will cost less to purchase but more to operate than a more expensive high efficiency model; this could cause the business to spend more to own and operate the "inexpensive" motor over the service life of the installation. To determine the true cost of ownership or to identify the option that has the best long-term impact on the corporate bottom line, you need to consider more than just the first costs; you must perform a more complete financial analysis of each investment alternative. A Life Cycle Cost Analysis (LCCA) is a methodology that considers all *relevant* economic consequences over a given period of time (or life cycle). This technique considers the present and future cash flows (both positive and negative) as well as the time value of money.

The basic LCCA process consists of two steps:

1. Identify *relevant* cash flows in terms of their quantity and time; and
2. Convert those amounts to a common frame of reference, usually a present worth or equivalent annual charge. Conversions are done with the aid of Time Value of Money conversion factors. These factors can be found in tables in engineering economy reference books (such as published by the Association for Facilities Engineering) and are programmed into financial calculators (e.g., the Hewlett Packard HP12C and the Texas Instruments Business Analyst II) and popular spreadsheet programs (such as EXCEL and Lotus 123).

Cash Flow Elements

As implied by its name, Life Cycle Cost Analysis seeks to include costs (and savings and/or income) that an initiative causes throughout its life. Candidates for inclusion can be categorized as follows:

- Initial Capital Investment
- Financing
- Operations and Maintenance
- Repair and Replacement
- Alteration and Improvement
- Salvage Value
- Revenue
- Taxes

Initial capital investment could include material, labor, shipping, engineering, project management/supervision, land, demolition of existing structures, permits, and inspection fees.

Financing costs could include loan origination fees and points, broker fees on bond issues, interest and principal paid to bond holders or loan issuers, and repayment of capital and interest to investors.

Operations and maintenance could include employee costs, energy, raw materials (if you are manufacturing a product), and contractor services. *Repair and replacement,* and *alteration and improvement* expenses could include staff labor, contractor services, and materials and parts. If the project components have any residual value at the end of the study life, then you need to consider its *salvage* value. Finally, your study needs to recognize any *revenue* that it may produce and any personal property, real estate, and income *taxes* that will have to be paid as a result of the implementation of the plan.

Note that in previous paragraphs the cash flow factors were frequently qualified by the words "relevant" and "could include." The nature and purpose of your study and certain individual circumstances will dictate which factors need to be addressed in your study. For example, when examining the economic impact of a new production line, all of the above factors are relevant to the decision-making process and need to be included. Otherwise you would be risking investing in an unprofitable project. On the other hand, when choosing between two options for replacing an irreparable chiller, you would ignore any common cost (such as the cost for removing the old unit), and include only those cost items or features that are different. Additionally, you would omit any costs that are too small to impact the results, that cannot be accurately quantified, or that are not available. (In the latter two instances it would be a good idea to estimate a reasonable range for the values and test to see if the maximum or minimum values cause the study outcome to change.) Finally, the nature of your business will enable you to eliminate some factors (e.g., taxes would not be considered by a non-profit enterprise).

Time Value of Money

All cash flows are classified in one of three categories: *Present, Future,* or *Annuity.* The Time Value of Money (or Engineering Economics) deals with the relationships between these three categories and their equivalence. For example, if you can earn 5% on your investments, receiving a payment of $1,000 today would be equivalent to receiving $1,050 one year from today. It is the equivalence relationships that enable you to use the conversion formulas and Time Value of Money tables to convert all cash flows to a common form and make valid comparisons between proposals. This process will be illustrated shortly in a couple of examples.

Conversions are performed at the "discount rate" determined by your finance department. It will be the interest rate on money borrowed, the corporate internal rate of return, or the amount the corporation can earn on external investments. In addition to the interest rate, the other key factor in the conversion process is the period of time covered by the study.

Life Cycle Cost Analysis Output/ Results

The results of a Life Cycle Cost Analysis (LCCA) can be described in a variety of units or terms, e.g., *Discounted Payback Period (DPP), Return on Investment (ROI), Present Worth of Expenditures (PWE), or Equivalent Annual Charges (EAC)*. The units that you will use to describe your project will probably be the ones with which the approving authority is most comfortable. With the exception of DPP, the superior project will always compare favorably regardless of the units used. DPP can be used, but it should never be used as the *sole* criteria for approving one proposal over another because there are some circumstances where DPP can be shorter for an option that otherwise is inferior to its competition. If Option "A" has a better ROI than Option "B," it will also have a better PWE and EAC. The bottom line is that your choice of units for describing your project will be driven by your target audience. Choose the terms that they are familiar with and expecting.

Examples

Sometimes selecting between options is easy...

I.	Option "A"	Option "B"
Initial Cost	$5,000	$8,000
Operating Expenses	$2,000/yr.	$2,500/yr.

Here, it is obvious that "A" is the superior plan since both first costs and ongoing operating expenditures are less. However, the choice would not be as clear if the Operating Expense values were reversed.

II.	Option "A"	Option "B"
Initial Cost	$5,000	$8,000
Operating Expenses	$2,500/yr.	$2,000/yr.

To identify the better plan, you would need to know the duration of the installation and your cost of money or discount rate. For the purpose of this example, let's use ten years and 8%, and solve for the present worth (PW) of the expenditures.

Option "A"

PW (Option "A") = PW (Initial Cost) + PW (Annual Costs)

We define the "initial cost" as occurring at the start of the project or in the "present" time frame. Therefore, the Present Worth of the initial cost is equal to the initial cost; no conversion is required. The Operating Expenses are expressed as an annual expenditure which therefore must be converted to a present value by the use of the following formula:

PW (Annual Costs) = (Op. Exp.) x (P/A) @ 8%, 10 Yrs.

Where (P/A) is the Present Worth of an Annuity factor that can be found in the accompanying Time Value of Money table. For this problem we will use the 8% table (see Appendix C) and use the P/A factor for 10 years (6.7101).

PW (Op. Exp.) = $2,500 x 6.7101 = $16,755.50

We now return to our original formula to solve the PW (Option "A"):

PW (Option "A") = PW (Initial Cost) + PW (Annual Costs)

= $5,000 + $16,755 = $21,775

which is the equivalent present worth of the expenditures associated with Option "A." Said another way, if you received $21,775 today and could earn 8% on money held for future use, you could fund the initial $5,000 investment and the $2,500/year operating expenses for ten years! Don't let the fact that your company wouldn't operate in this manner throw you off course. Remember that the purpose of the conversion is to get all cost elements into a common format — present, future, or annual costs — for the purpose of comparison. It does not indicate how you would actually operate.

Option "B"

PW (Option "B") = PW (Initial Cost) + PW (Annual Costs)

= $8,000 + $2,000 (P/A) @ 8%, 10 Yrs.

= $8,000 + $13,420 = $21,420

Now that we have both Options "A" and "B" expressed in a common format, Present Worth, we can compare the two and identify the one that will cost the least over the life of the installation.

	Option "A"	Option "B"
Present Worth	$21,775	$21,420

When the closeness of the values makes the decision a virtual toss-up, you will need to consider other factors, such as inflation. If you were doing a revenue requirements type of analysis, it would be essential that all costs be fully accounted for, and you would have to calculate the impact of inflation. However, in simple comparison situations, it often sufficient to simply apply logic. In this example, we can be pretty confident that operating expenses will rise over time. However if we assumed that inflation will affect each option in the same way, we can logically conclude that the Option "A" expenses will increase more than those for "B." Since "B" already has the economic advantage, inflation will only serve to increase that advantage. Here no further calculations would be necessary. Accounting for inflation will be illustrated in the next example.

Finally, even when there is a clear "winner," your choice may be influenced by the availability of capital. If the budget is tight, you may not be able to fund the more economical option even if it is clearly a far superior choice.

The previous example is extremely simplistic, dealing only with initial and annual operating costs, and expressing those costs in "lump sum" terms. In the following example, we will add a third type of expenditure (a future cost), break the other cost elements into their component parts, and partially take inflation into account.

III. The furnace serving the administrative offices in your plant needs to be replaced. You have narrowed the decision down to options (A), an 80% efficient unit, and (B), a more expensive, 95% efficient unit. The estimated costs and study factors are as follows:

	"A"	"B"
Hardware Cost	$1,500	$2,200
Installation Labor (using in-house staff)	300	300
Annual Servicing/Tune-up	50	75
Annual Energy Use	1,200	1,000
Igniter Replacement (every 10 years; option "A" only)	120*	—
Vent Damper Motor Replacement (every 12 years. option "B" only)	—	200*

General Assumptions:

Study period = 20 years

Investment rate = 8%

Inflation rate = 4%

*Based on industry experience, you expect to have to replace these components at specific times in the future. While you don't know what the future costs will be, you do know what the work would cost if done today. To improve the accuracy of our study, we will estimate those future costs by applying an inflation factor.

As with the previous example, we will convert all expenditures to their equivalent present worth. The PW for either option is:

PW (Option) = Initial Costs + PW (Annual Costs) + PW (Future Costs)

Option "A"

Initial Costs = Hardware + Installation Labor

= $1,500 + $300 = **$1,800**

PW (Annual Costs) = PW (Annual Service) + PW (Energy)

= $50 x (P/A) @ 8%, 20 years + $1,200 (P/A) @ 8%, 20 years

= $50 x (9.8182) + $1,200 x (9.8182) = **$12,273**

So far the process has been virtually identical to that used in Example II. You have probably noticed that the annual cost elements could have been combined before multiplying by the conversion factor, which raises the question, "Why separate the costs?" The purpose of separating the cost components is to allow you to see that all relevant factors have been included, and that the estimate for each is reasonable. After documenting the various components, it is perfectly acceptable to combine them numerically when performing this conversion.

The final element in our study is the cost for the future replacement of the igniter. To make our results a little more accurate, we will use the Time Value of Money tables to estimate what the actual charges will be in the future, then convert those future costs to their equivalent present value. We know that it would cost $125 to replace the igniter we did it today. Based on historical trends for this type of service, we estimate that this cost will increase approximately 4% per year. To calculate a future value from a present amount:

FW (Replacement) = Present Cost x (F/P) @ 4%, 10 years

= $125 x (1.480) = $185

At this point we are halfway home. We have just calculated what we expect to pay for this work when it is done in ten years. But remember that our goal is to convert all costs to present value, which means we now must convert these future dollars back to their present equivalent. Keep in mind that we are dealing with a hypothetical situation in which we are calculating the amount we would have to invest today to have the funds necessary to pay this $185 bill when it comes due in ten years. Since we are dealing with the amount we would invest, we must use the investment rate (8%) when we do the conversion, not the inflation value!

PW (Replacement) = Future Cost x (P/F) @ 8%, 10 years

= $185 x (.4632) = $86

We now have all of the costs to plug into our original formula:

PW (Option "A")   = Initial Costs + PW (Annual Costs) + PW (Future Costs)

= $1,800 + $12,723 + $86 = $14,609

Following the pattern used in solving for "A", solving for "B" is quite easy...

Option "B"

Initial Costs   = Hardware + Installation Labor

= $2,200 + $300 = $2,500

PW (Annual Costs) = PW (Annual Service) + PW (Energy)

= $75 x (P/A) 8%, 20 years + $1,000 (P/A) @ 8%, 20 yrs.

= ($75 + $1,000) x (9.8182) = $10,555

FW (Replacement) = Present Cost x (F/P) @ 4%, 12 yrs.

= $200 x (1.601) = $320

PW (Replacement) = Future Cost x (P/F) @ 8%, 12 yrs

= $320 x (.3971) = $127

PW (Option "B")   = Initial Costs + PW (Annual Costs) = PW (Future Costs)

= $2,500 + $10,555 + $127 = $13,182

With all costs converted to a common base — Present Worth — it is easy to compare the two choices and see that Option "B" costs less than "A."

	Option "A"	Option "B"
Present Worth	$14,609	$13,182

Besides Present Worth, it is fairly common to convert all cost factors to an *Equivalent Annual Charge*. Many managers are more comfortable with this frame of reference because it indicates how much revenue must be collected to pay the actual and equivalent annual expenses associated with the project. Using the Time Value of Money tables we can convert the Present Worth results of Example III to an Equivalent Annual Charge.

$$EAC = PW \times (A/P) @ i\%, \text{n years}$$

EAC (Option "A")   = PW ("A") x (A/P) @ 8%, 20 years

= \$14,609 x (0.1019) = \$1,489

EAC (Option "B")   = PW ("B") x (A/P) @ 8%, 20 years

= \$13,182 x (0.1019) = \$1,343

As expected, Option "B" is still less expensive than Option "A," with ownership and operating cost being \$145/year less.

Appendix C of this book is a collection of four tables that allow you to calculate Present Worth.

Summary

Effective cost management requires carefully prepared feasibility studies, a cost estimate, and a responsive cost monitoring and reporting system based on an established work breakdown structure. The design and budget stage of the cost management process may include a Life Cycle Cost Analysis. The relative simplicity or complexity of the analysis will depend on your situation; there is no single approach. For example, one organization may be chiefly interested in payback, while another is in a position to commit to a long-term capital expenditure. An established history of good cost analysis and control procedures may be the factor that tips the scale in getting your plans approved over competing corporate investment alternatives.

Chapter 7

Schedule Planning and Control

One way that an organization or individual can gain and sustain a competitive advantage is to produce the greatest amount of high quality work in the least amount of time. Many organizations can dramatically increase their productivity by improving their ability to plan and control resources. The primary objective of scheduling is to make the most efficient use of available resources: time, money, manpower, materials and equipment.

One measure of project performance is the time period required to complete the project's activities. Many of the costs involved in completing a project are time-related, such as labor, supervision, equipment, overhead and financing. The many factors that can affect scheduling during a project can also affect cost. They cannot be separated. The only difference between estimating and scheduling is that one is cost-focused and the other is time-focused. The work breakdown structure is the common denominator. Based on the work breakdown structure for the project, activity durations are estimated. The durations may be based on historical data, published labor productivity standards, consultant and subcontractor-supplied information, and resource availability. The schedule should include all relevant activities including programming, design and engineering, procurement, permitting, construction, and start-up activities. There is a need to schedule blocks of time for coordination and quality control meetings, as well as final review and sign-offs. Failing to allow for these activities can be a major source of change orders and project delays. (See Chapter 5 for more on the work breakdown structure.)

Project Scheduling Methods

Over the years many different scheduling methods have been devised to control all kinds of projects. Milestone charts, bar charts, and critical path method (CPM) are the most popular scheduling methods used for planning and controlling facilities construction and maintenance projects. These tools provide the overall planning, scheduling, and control needed to sequence operations properly and to allocate resources efficiently.

No one scheduling method is appropriate for every project. All scheduling methods have advantages and disadvantages, and each project must be scheduled using a method that is best suited to the particular project. The scheduling method selected by the project manager should take into consideration the following:

1. Size and complexity of the project
2. Scope of work
3. Number of activities and disciplines involved
4. Necessity and frequency of updates and revisions
5. Expectations of those who will be using the schedule
6. Budgeted time and cost for preparing the schedule

Milestone Charts

The milestone chart is probably the simplest scheduling method. Basically, the chart consists of a list of project events or milestones in the general order in which they are planned to be accomplished. Milestones are simply a point in time identifying significant accomplishments in the progression of a project. They have no time duration. Some examples of construction project milestones might include: appropriation of project funding, issuance of a notice to proceed from a project owner to a contractor, completion of construction documents, completion of the foundation, issuance of a certificate of substantial completion by the project designer, or receipt of a construction permit or occupancy certificate from the local municipality.

To avoid excessive detail when preparing a milestone chart, list the target completion dates for key project activities only. Frequently, the milestone chart also identifies who is responsible for performance of the tasks leading up to each milestone. The major advantages of a milestone chart are ease of preparation and emphasis on summarized target completion dates. The facility manager's best application for a milestone chart is during the early planning phase of a capital development project. It can also be used on projects of short duration with few participants and little interrelationship between activities. A milestone chart is often incorporated into a contract between the project owner and design professional, and in the contract between owner and contractor. The milestone chart summarizes and depicts significant work packages within major work categories. For example, it can be used to communicate deadlines for delivery of major design items in a construction project.

A major disadvantage of the milestone chart is that it shows only events or target completion dates. This results in uncertainty about when each project activity should begin. A milestone chart does not communicate activity durations and interrelationships. This omission renders a milestone chart ineffective as a project management tool for optimizing and controlling timing cycles of activities to minimize direct and indirect costs. See Figure 7.1 for a sample milestone chart.

Bar Charts

The work-versus-time schedules developed in the early 1900s by Henry Gantt and Frederick Taylor are considered the first scientific approach to scheduling. These Gantt charts were originally developed for use in scheduling manufacturing operations. However, they have become readily accepted as a tool for scheduling project activities and recording progress.

A Gantt Chart, commonly known as a *bar chart*, graphically describes activities on a work-versus-time scale. The chart illustrates the planned start and completion date for the various project activities.

A bar chart is generally organized so that the work activities are identified in a column at the left side of the chart. For example, an activity on a bar chart for the design phase of a building construction project may be "complete thermal load calculations." Figure 7.2 is a sample bar chart for a fast-track print center remodeling project.

A bar representing the duration of time for each activity is drawn between the scheduled start and completion time for each activity along a horizontal line. At the top of the chart is a horizontal time scale. The difference between planned and actual progress can also be reported on a bar chart. The bar chart assumes a direct linear relationship between progress of an activity as a percentage of

Milestone Chart

Activity Identification	Activity Description	Completion Date
100	Predesign	February 17
200	Site Analysis	March 31
300	Schematic Design	May 30
400	Design Development	July 25
500	Construction Documentation	October 10
600	Bidding and Negotiation	October 24
700	Construction Phase	October 31

ɡure 7.1

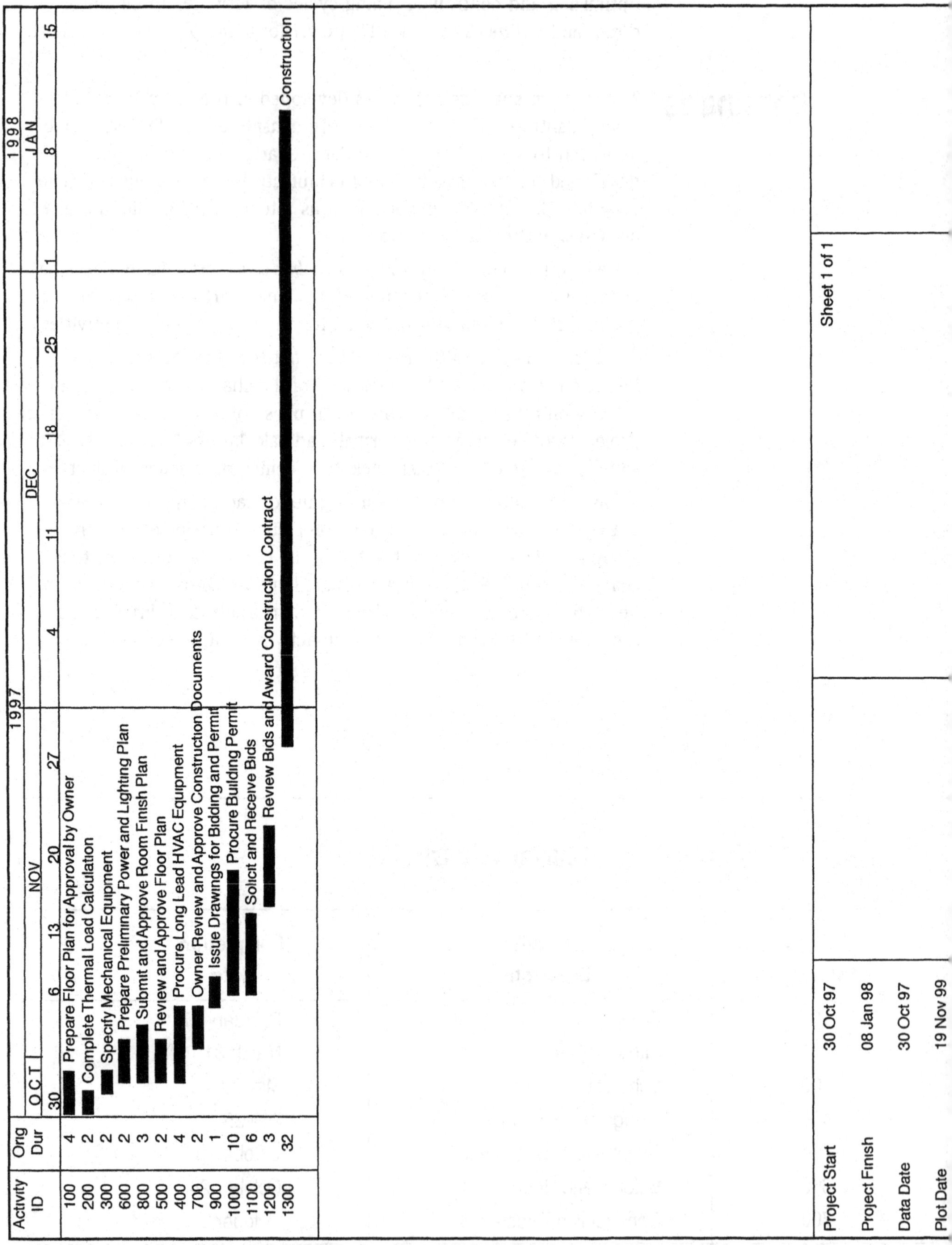

Figure 7.2 Sample Print Center Remodeling Project—Houston, Texas

Activity ID	Orig Dur
100	4
200	2
300	2
600	2
800	3
500	2
400	4
700	2
900	1
1000	10
1100	6
1200	3
1300	32

Activities (bar labels):
- Prepare Floor Plan for Approval by Owner
- Complete Thermal Load Calculation
- Specify Mechanical Equipment
- Prepare Preliminary Power and Lighting Plan
- Submit and Approve Room Finish Plan
- Review and Approve Floor Plan
- Procure Long Lead HVAC Equipment
- Owner Review and Approve Construction Documents
- Issue Drawings for Bidding and Permit
- Procure Building Permit
- Solicit and Receive Bids
- Review Bids and Award Construction Contract
- Construction

Project Start	30 Oct 97
Project Finish	08 Jan 98
Data Date	30 Oct 97
Plot Date	19 Nov 99

Sheet 1 of 1

162

planned progress. Because the bar chart can be quickly and easily interpreted, it can be useful for presenting summarized information to senior management. It can also be used to identify and communicate activity durations and time allotments to project personnel, including tradespeople and maintenance personnel.

To report progress, an open bar is shaded in as the work progresses, to identify work completed, and not just elapsed time. See Figure 7.3 for an example involving a remodeling project.

Advantages: Bar charts have a number of advantages over other scheduling methods.

1. Ease of preparation
2. A visual summary that is easily understood
3. A good communication tool for presenting uncomplicated plans (plans with a limited number of activities and dependencies)

Disadvantages: The limitations of traditional bar charts must be understood. Among them:

1. A bar chart shows time frame, but does not show relationships and dependencies among project activities.
2. A bar chart becomes cumbersome to use as the number of activities increase.

Since a bar chart does not communicate sequence constraints, it cannot readily be used to forecast the effects of activity changes (delays, additions, or deletions) to the scheduled completion date. Failure to obtain a delivery on time, a schedule slippage due to extra work or unforeseen site conditions, or a change in the scope of the work may change the length of one or more bars on the schedule, but the effect on other activities will not be apparent.

Bar charts remain an important planning and scheduling tool. However, their use as a control tool and their limitations should be understood for appropriate application.

Network Scheduling

This scheduling method involves connecting activities in a series of chains, or networks, to communicate the necessary sequence and relationships between activities as determined by the planner to achieve project objectives. The network schedule is a flow chart of the work breakdown structure. The timing and sequence of activities in a network may be dictated by technical constraints (such as completing foundation work prior to the construction of the superstructure); by the cost of and availability of resources (such as excavating for site sewer and plumbing while equipment is on location for excavating the foundation); by safety (such as delaying work until an overhead activity is finished in order to avoid a potentially unsafe condition); and by common sense and project owner requirements. Activities may also be scheduled as they are because it is the politically correct thing to do.

SUBURBAN HOSPITAL SCHEDULE

Task Name	Duration	Start	End	Responsible
Receive Signed Contract	1.00d	14/Feb	14/Feb	SMS
Order Equipment	1.00d	14/Feb	14/Feb	SMS
Issue Subcontractor PO's	1.00d	15/Feb	15/Feb	SMS
Receive Business Licenses for municipality from all Subs	8.00d	15/Feb	25/Feb	SMS
Receive Shop Drawings & Submittals	1.00d	17/Feb	17/Feb	Custom Interiors
Submit Permit Application from Village	8.00d	17/Feb	01/Mar	SMS
Submit Structural Shop Drawings to Engineer	20.00d	23/Feb	22/Mar	Stl Cont./SMS
Structural Fabrication/Installation	28.00d	15/Mar	21/Apr	Phoenix Welding
Erect Siding	8.00d	22/Apr	03/May	Custom Interiors
Roofing of New Structure	5.00d	04/May	10/May	Roofing Contractor
Remove Existing Roof	3.00d	18/Apr	20/Apr	Custom Interiors
Pour/Cure Housekeeping Pad	4.00d	25/Mar	30/Mar	Concrete Co
Masonry Work & Shot Blasting	5.00d	25/Apr	29/Apr	Custom Interiors
Fireproofing	5.00d	02/May	06/May	Custom Interiors
Bring Piping Out to Parapet	20.00d	03/Mar	30/Mar	SMS
Receive and Set AHU	1.00d	30/Mar	30/Mar	SMS/Vent. Cont
Penthouse Piping	20.00d	21/Apr	18/May	SMS
Insulate Piping & Ductwork	20.00d	09/May	06/Jun	Insulation Contractor
Electrical	10.00d	09/May	20/May	Electrical Contractor
Sheet Metal Fabrication	21.00d	15/Mar	12/Apr	Vent. Contractor
Sheet Metal Installation	21.00d	15/Apr	13/May	Vent. Contractor
Final Sheet Metal Tie-In	1.00d	01/Jun	01/Jun	Vent. Contractor
Temperature Control Installation	20.00d	26/Apr	23/May	Controls Contractor
Demo Existing AHU	15.00d	01/Jun	21/Jun	Vent. Contractor
Start-Up New AHU	1.00d	26/May	26/May	SMS
Punch List	20.00d	01/Jun	28/Jun	SMS

Figure 7.3

Printed 30/Mar
Page 1

164

Critical Path Method (CPM)

The critical path method was developed in the 1950s by the Dupont Corporation to reduce the amount of time and money required to construct their chemical plants. Today, CPM is used for:

- Scheduling construction, production, and maintenance projects;
- Introducing new products into the marketplace; and
- Improving work flow in both manufacturing and service operations

A CPM schedule is a type of network diagram that effectively links all activities of a project. It shows the relationship and dependencies among project activities. There is at least one path of activities through the network from start to finish that controls the overall duration of the project. This is the critical path. The sum of activity durations along the critical path of the network shows the shortest amount of time possible to complete the project. The critical path is the sequence of activities through the network which controls the project duration. Figure 7.4 is a sample CPM schedule for the print center remodeling project (previously shown in bar chart format in Figure 7.2).

Critical Activities

If any activity on the critical path takes longer than its estimated duration, the overall project completion date will be delayed accordingly. An activity that is not on the critical path has more time available for completion than the time actually required to perform the task. This extra time is called "float." CPM diagrams show how much time activities not on the critical path can be delayed without affecting the overall project completion date. Noncritical activities may become critical if the time required to complete them exceeds their allotted time plus float.

The critical path is calculated to reflect when each activity must be completed in order to finish a project as early as possible. Another big reason to use a CPM schedule is to allow management to exercise its options when the availability of resources is limited. For example, if a deadline for a project is in jeopardy, work on a noncritical activity may be delayed to concentrate available labor on a critical one. Work can be delayed on any noncritical activity, as long as its float time is not exceeded. This process is referred to as *resource leveling* and will be addressed later in this chapter.

The total project duration can be shortened only by reducing the amount of time required to complete critical activities. For example, if the curing of concrete is on the critical path, a decision may be made to use a high-early strength concrete to shorten the amount of time required to complete that activity. If the critical activity is labor-intensive, one approach might be to add a second shift or to outsource some or all of the work.

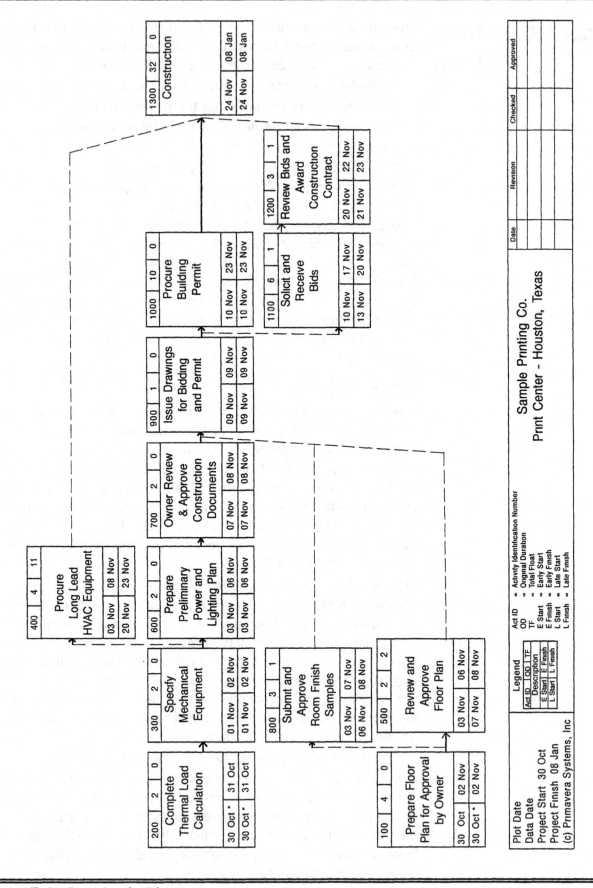

Figure 7.4   Pure Logic Diagram

CPM assigns a fixed time duration to each activity. Usually activities are identified by writing a few descriptive words, often starting with a verb. Once the project activities have been listed with an estimated fixed time duration for each, their sequence and relationships can be defined.

The activities and their relationships can be listed in a table with the corresponding dependencies of each. Activities that must be completed before another activity can be started are called *predecessor activities*. An activity whose start is dependent on other activities that must logically precede it is called a *successor activity*. *Dependencies* are the activities which immediately precede an activity. When determining the sequence and relationships between the activities that comprise a project, consider which ones may be performed simultaneously and which of the immediately preceding activities *must* be completed before a succeeding activity can begin. Figure 7.5 is a list of activities with their corresponding dependencies and assigned durations. Refer back to Figure 7.4 and note how the activity numbers appear in the nodes or boxes.

Two types of CPM scheduling methods are used to calculate the critical path: *Activity-On-Arrow (AOA)* and *Activity-On-Node (AON)*. While the two methods may appear, at first glance, to be quite different, they are basically the same. The AOA network model was the original network scheduling method developed. It is sometimes referred to as the *arrow diagramming method*.

The Activity-On-Node approach to CPM is less complicated and is becoming more widely used than the Activity-On-Arrow diagram for both manual calculations and computer scheduling. Therefore, the Activity-On-Node method will be used here to demonstrate the CPM scheduling method. The node or box represents the activity. A line called a *link* is used to connect two activities.

Activities were originally represented by circles and connecting arrows to show the logical relationship between them on AON network schedules. Today, activities are often shown as boxes. See Figure 7.6 for examples of AON and AOA diagrams.

For manual CPM calculations activity relationships are often restricted to finish-to-start relationships where a succeeding activity cannot start until a preceding one has been completed. However, most computer CPM scheduling programs better reflect reality by allowing for gaps between activities and activity overlaps. Figure 7.7 shows some possible precedence relationships.

Any unit of time can be used in performing network schedule calculations. As with all scheduling methods, the only requirement is that you must select a consistent unit of time for expressing the duration of all activities in the network. Schedules may be expressed in units of years, months, weeks, days, hours, or even minutes. Schedules expressed in hours can be useful in planning process shut-downs to minimize production downtime due to maintenance and

Sample CPM Organizational Sheet

Date: _____

Title: Building Remodeling Project. New Print Center _____

Project Location Houston, TX _____

Number	Activity I.D.	Activity Description	Activity Duration (Calendar Days)	Immediate Predecessor Activity (Dependencies)
100	A	Prepare final floor plan for approval by owner	4	—
200	B	Complete thermal load calculations	2	—
300	C	Specify mechanical equipment	2	200
400	D	Procure long lead HVAC equipment	4	300
500	E	Review and approve floor plan	2	100
600	F	Prepare preliminary power and lighting plan	2	300
700	G	Owner review and approve construction documents	2	600
800	H	Submit and approve room finish samples	3	100
900	I	Issue drawings for bidding and permit	1	500;700,800
1000	J	Procure building permit	10	900
1100	K	Solicit and receive bids	6	900
1200	L	Review bids and award construction contract	3	1100
1300	M	Construction	32	400,1000;1200

Figure 7.5

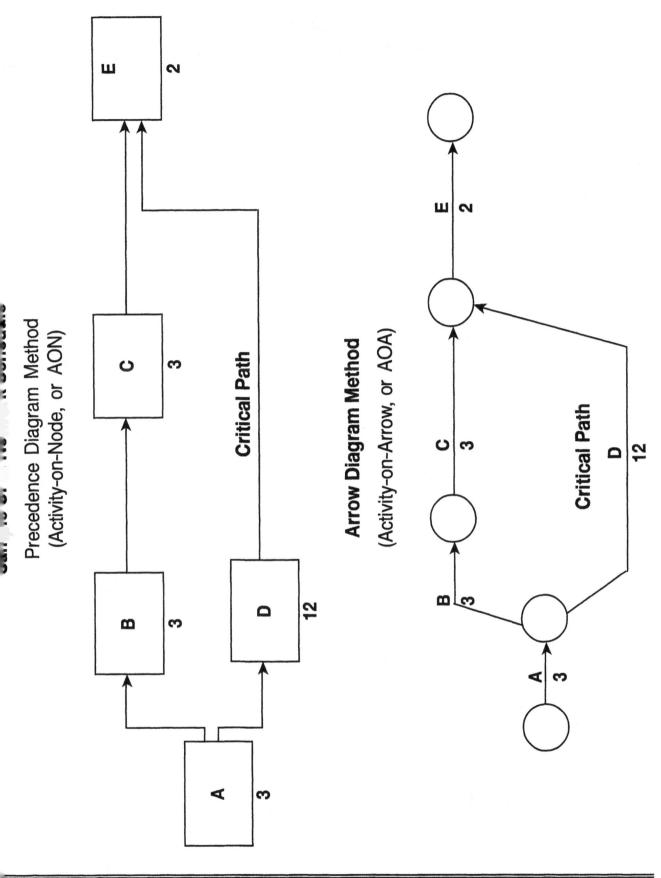

Samples of the Project Schedule

Precedence Diagram Method
(Activity-on-Node, or AON)

Critical Path

Arrow Diagram Method
(Activity-on-Arrow, or AOA)

Critical Path

Figure 7.6

169

Precedence
Relationships Between Activities

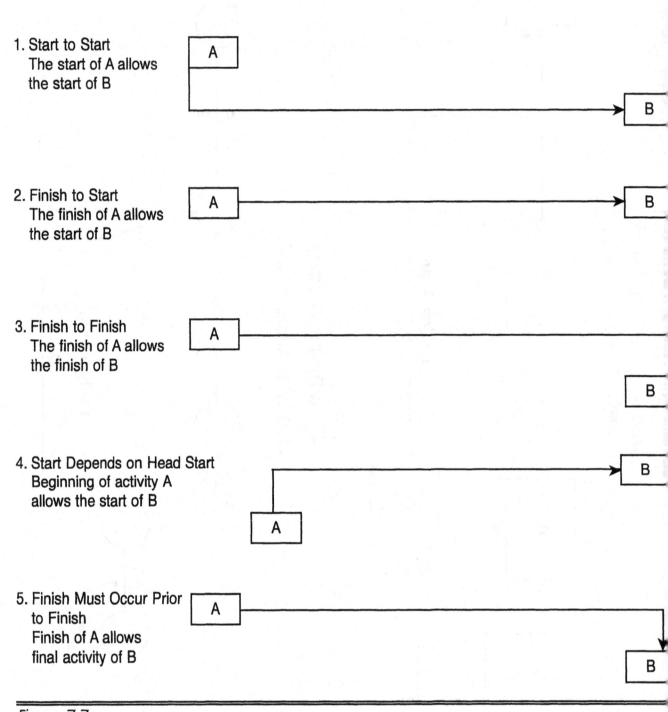

1. Start to Start
   The start of A allows
   the start of B

2. Finish to Start
   The finish of A allows
   the start of B

3. Finish to Finish
   The finish of A allows
   the finish of B

4. Start Depends on Head Start
   Beginning of activity A
   allows the start of B

5. Finish Must Occur Prior
   to Finish
   Finish of A allows
   final activity of B

Figure 7.7

repair-type projects. Schedules expressed in units of weeks may be suitable for large, long-duration projects. Time units expressed in days are most commonly used for planning and managing construction projects. The time unit used to demonstrate CPM scheduling methods in this chapter will be days.

Several calculations are required for preparing a CPM schedule: the forward pass, the backward pass, total float, and free float. Figure 7.8a is an example of a manually-generated CPM activity-on-node precedence network for the print center remodeling project (previously shown in bar chart and computer-generated network diagram formats in Figures 7.2 and 7.4). Figure 7.8b is a calculation of total float and free float in a column listing. The following section explains how to calculate the CPM schedule.

The Forward Pass

The first calculation is the forward pass. The purpose of performing a forward pass is to calculate the early start (ES) and early finish (EF) time for each network activity. The ES is the earliest time each network activity can start, allowing for the time required to complete preceding activities. The EF of an activity is the earliest time that activity can finish given the network logic and activity duration. The earliest an activity can finish is the early start date of that activity, plus that activity's duration. Completion of the forward pass provides the earliest date the project can finish based on the sequence of activities and activity durations.

Begin the forward pass with the first activity in the network. In performing the forward pass, all activities are assumed to start as early as possible after all their immediately preceding activities have been completed. Activities that are immediately preceded by the same activity must have the same early start date. The earliest that a successor activity with more than one immediate predecessor can start is determined by the early finish date of the last predecessor to finish.

> ES = the latest early finish date of a given activity's immediate preceding activities.

The ES of the initial activity in a network must have its early start assigned by the scheduler. The assigned early start is usually zero, but may be some other desired point in time if the schedule is coordinated with other projects in the organization.

> EF = ES + Duration

The shortest possible time for completing the total project is determined by the early finish time of the last activity in the network.

The Backward Pass

The next step is to work backward through the network. The backward pass determines the latest time each activity in the network can start (late start date, or LSD, or LS) and finish (late finish date, or

Sample Manual Calculation for Activity-on-Node CPM Network Schedule
Print Center Remodeling Project

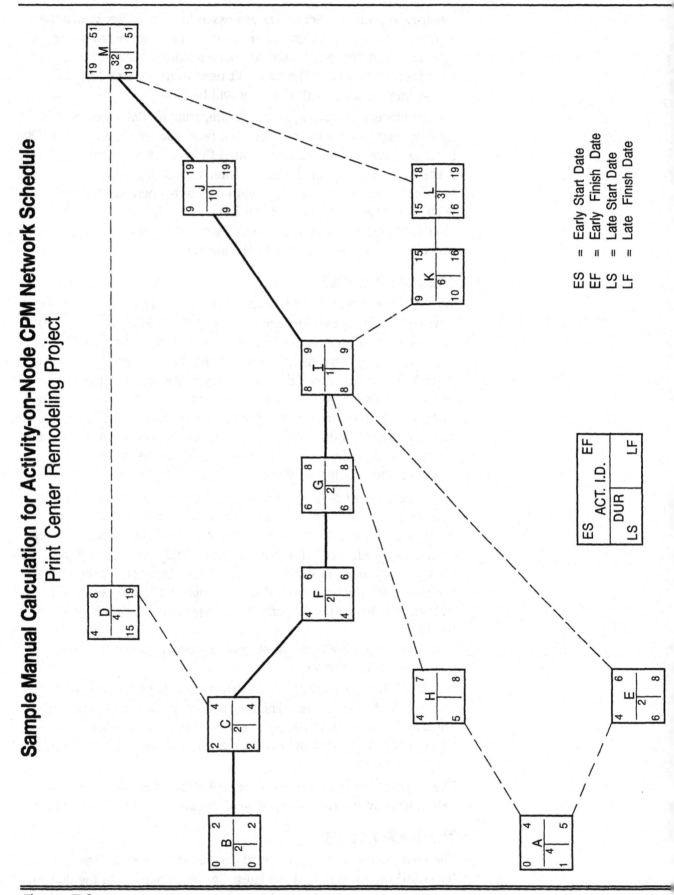

ES = Early Start Date
EF = Early Finish Date
LS = Late Start Date
LF = Late Finish Date

Figure 7.8a

Calculation of Total Float and Free Float

Activity	LFD	–	EFD	=	Total Float
A	5	–	4	=	1
B	2	–	2	=	0
C	4	–	4	=	0
D	19	–	8	=	11
E	8	–	6	=	2
F	6	–	6	=	0
G	8	–	8	=	0
H	8	–	7	=	1
I	9	–	9	=	0
J	19	–	19	=	0
K	16	–	15	=	1
L	19	–	18	=	1
M	51	–	51	=	0

Activity	Successor Min. ESD	–	EFD	=	Free Float
A	4	–	4	=	0
B	2	–	2	=	0
C	4	–	4	=	0
D	19	–	8	=	11
E	8	–	6	=	2
F	6	–	6	=	0
G	6	–	6	=	0
H	8	–	7	=	1
I	9	–	9	=	0
J	19	–	19	=	0
K	15	–	15	=	0
L	19	–	18	=	1
M					

Critical Path =
2 + 2 + 2 + 2 + 1 + 10 + 32 = 51 Days
(determined by the sum of the activity
durations during the critical path)

LFD = Late Finish Date
EFD = Early Finish Date
ESD = Early Start Date

gure 7.8b

LFD, or LF) without delaying the minimum total project duration as calculated by the forward pass.

LSD = LFD – Duration

LFD = earliest LSD of all successor activities

The amount of float, sometimes called *slack*, is calculated by manipulating the values determined from the forward and backward passes. Two types of float calculated for each activity in the network schedule are *total float* and *free float*.

Total Float (TF) is the time span in which an activity may be delayed without delaying the completion of the project. Total float is the number of days between either the early start date and late start date or the early finish date and late finish date. Activities with an early start date equal to the late start date have zero float and are therefore critical activities. An activity is critical if its total float is equal to zero. Any continuous path from start to finish of a network schedule that contains activities with a total float of zero is a critical path.

TF = LSD – ESD or

TF = LFD – EFD

Free Float (FF) is the time span in which the finish of an activity may be delayed, without delaying the early start of any succeeding activity. Boundaries of this time span are the early start date of any immediately following activities and the early finish date of the activity itself.

FF = the earliest ESD (early start date) of succeeding activities – EFD (early finish date)

Total float and free float are commonly used.

Remember, there is at least one path through the project network, from start to finish, that controls the project length. It is called the critical path. It is the longest path of activities through the network, but actually represents the minimum time required to complete the overall project. When you add the duration of time along the critical path of the network, you can compute the minimum time required to do the project. The critical path is often shown by a solid or heavy line in the diagram. If a critical path activity is delayed, the project end date will be delayed by an equivalent amount of time. If an activity is not on a critical path, it has float. Activities on the critical path have no float. Early and late start times are the same for activities on the critical path. For activities not on the critical path, the early start date and early finish date are different from the late start date and late finish date.

Advantages and Disadvantages

CPM network schedules have these advantages over bar and milestone charts:

1. They show interrelationships among project activities; how completion of one activity affects completion of another; and, ultimately, the effect on the overall completion date.
2. They show the shortest possible time required to complete the project.
3. They enable effective management decision-making relevant to time and money by showing which project phases are critical.
4. They can be easily updated with computers.
5. They may be used to show completion and percent of completion dates so project owners can determine when progress payments are due and identify various cash requirements.

CPM network schedules have these disadvantages:

1. They are often time-consuming and costly to prepare.
2. They can be tough to communicate. However, information gained from a CPM can be abbreviated to other report formats.

Balancing Project Time and Cost Relationships (Resource Leveling)

Frequently, managers are confronted with the need to assign labor overtime or additional equipment to a project in order to meet a schedule deadline. This increases the overall project cost. The decision to reduce the project duration must be based on an analysis of the trade-off between time and cost. Projects are driven by resource cost and availability, as well as the project delivery deadline. Figure 7.9 is a histogram showing an imbalanced use of resources, whereby the senior designer of the composite crew (architect, electrical engineer, mechanical engineer, and senior designer) are overutilized, based on a normal, maximum eight-hour working day during the week of October 30th. Other weeks, the senior designer is underutilized. (In fact, none of the crew members is used at all the week of November 13th.) In Figure 7.10, the schedule has been leveled to better balance resource utilization. Figure 7.11 is a revised CPM schedule that shows the changes brought about by leveling. Note that the overall project duration has been extended by two days from the original. Various resource configurations can be calculated on the computer in an attempt to arrive at the optimum time/cost relationship. Resource leveling utilizes float time from noncritical activities to balance resources that were previously over- or underscheduled.

Summary

Facility managers must prioritize projects and optimize available time in order to cope with the pressure of increasing customer demands. Shorter product life cycles require accelerated delivery of new facilities and installation of required technologies. The introduction of Just-In-Time manufacturing requires a reduction in the cycle times of maintenance projects. To meet these demands, the facility manager must have a good understanding of the basic scheduling tools, and when to use each for best advantage. These schedules are an essential tool, not only for controlling the project, but for communicating with upper management and the project team.

The milestone chart serves as a road map for projects of relatively long duration. It is easily understood by all management levels, and provides adequate detail for long-range planning. Bar charts offer more detail, but still convey the information in a readable format, to both upper management and field personnel. The critical path method is the most precise scheduling tool. Its strength lies in its communication of the sequence and interdependencies of different activities in the complete project process. This format identifies key items that, if delayed, will extend the project duration, and thus the cost. The down side of CPM is that it is not readily understood by those outside of construction management, industrial engineering, or operations management.

No one scheduling method is always appropriate. All have been used successfully by project planners. Proper application of the appropriate scheduling method enhances communications and offers the best opportunity for reducing project delivery time and cost.

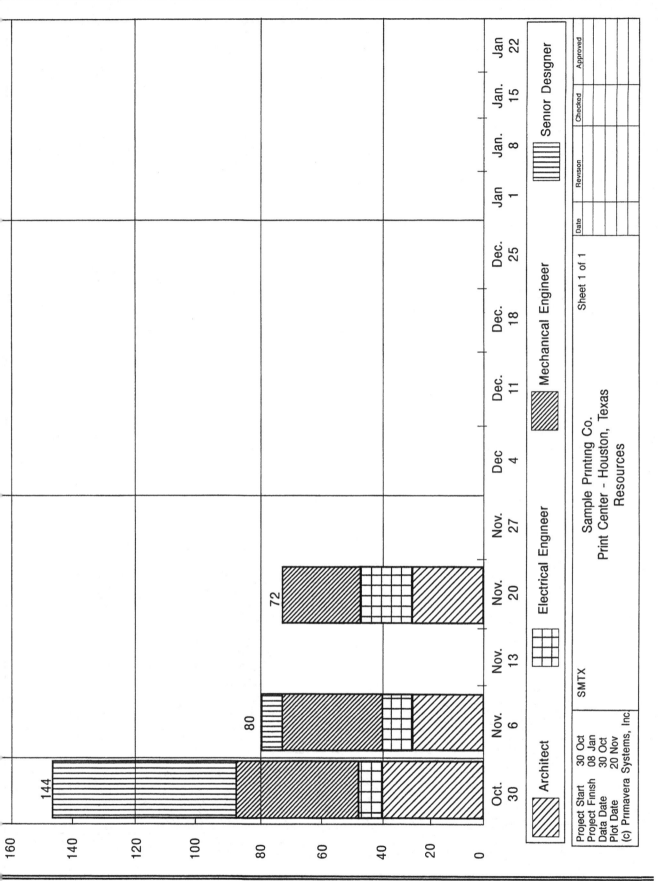

Figure 7.9 Resource Histogram (before leveling)

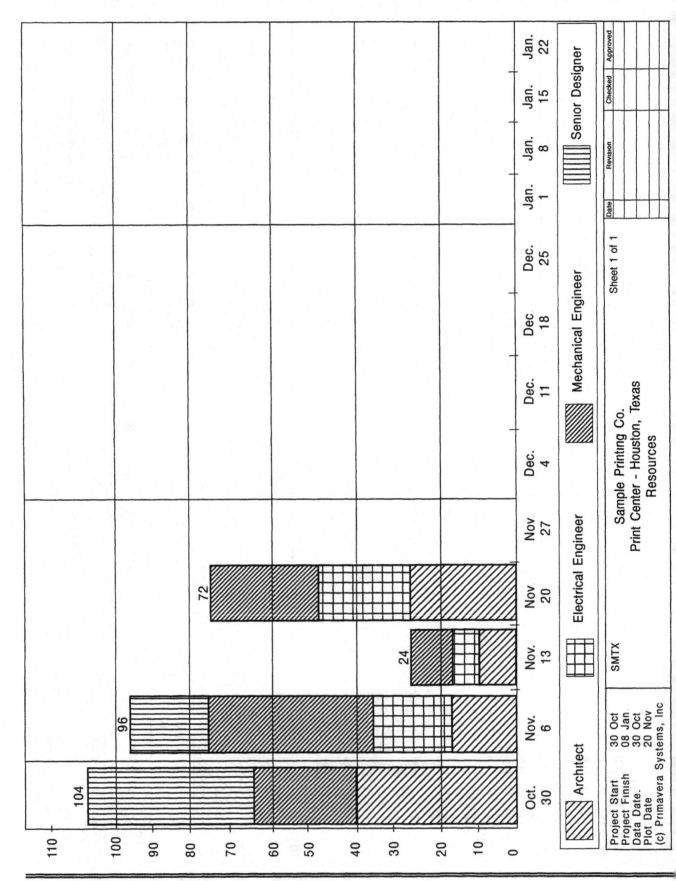

Figure 7.10 Resource Histogram (after leveling)

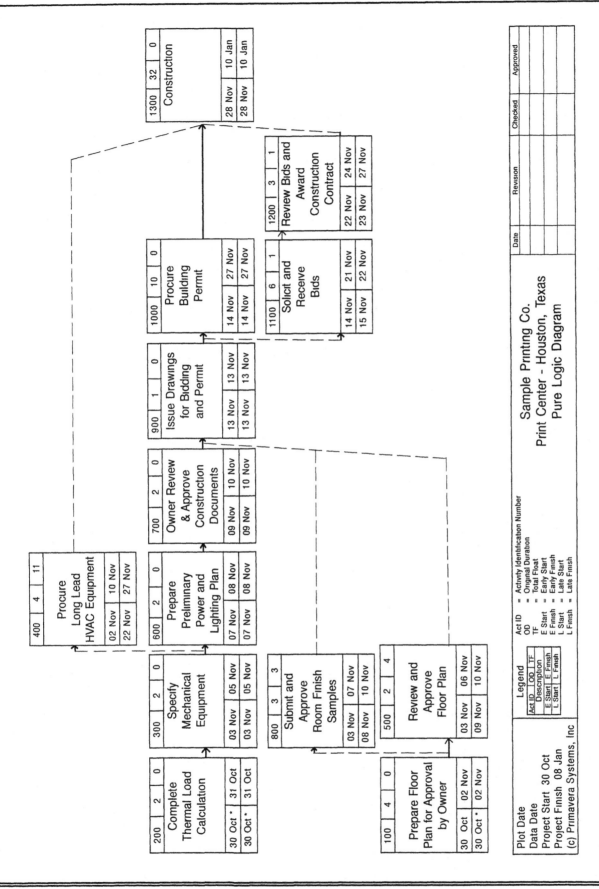

Figure 7.11   CPM Schedule (after leveling)

Chapter 8

Contracting and Procurement Methods

Developing a project can be difficult, time-consuming, expensive, and risky. There are numerous barriers to project development, including technical, financial, legal, environmental, and political. To overcome these barriers, you need to involve project personnel representing various disciplines. For example, the project may require the participation and collaboration of facility programmers, architects, engineers, construction managers, project managers, contractors, maintenance technicians, environmental and technical consultants, and government or regulatory agencies, as well as financial and legal experts. The ensuing roles, relationships and responsibilities need to be defined in the form of a written contractual agreement so that the parties work effectively as a team. With this arrangement in place, the probability of a successful project outcome is greatly increased.

Three basic project delivery methods are used in the United States: (1) general contractor, (2) design/build, and (3) construction management.

General Contractor

Using detailed and complete construction drawings and specifications, the owner either negotiates a contract with a specific general contractor or obtains competitive bids from two or more general contracting firms to perform only the construction. Under this arrangement, the project owner contracts separately with a design professional to provide the construction documents. General contractors normally perform some of the work with their own forces and award the rest to subcontractors. For example, structural steel, roofing, earthwork, concrete, mechanical, electrical, and other subcontractor trades are responsible to the general contractor for completion of their portions of the work, while the general contractor remains responsible to the owner for completion of all construction. On conventional building construction projects, approximately 80-90 percent of the general contractor's work is typically performed by subcontractors. Most owners prefer that the general contractor sign contract for a lump sum fixed amount.

Drawings and specifications must be completed and bids obtained from general contractors before a contract can be awarded and construction can begin. This results in an unnecessarily long project completion time and potentially higher cost. The absence of the contractor's constructability input during the design phase may also result in higher costs. Problems can also occur in the contractor's understanding of the construction documents provided by the designer. Issues in question may be interpreted in ways that result in greater expense to the owner. Furthermore, it costs more money and takes more time to make design changes in an effort to reduce construction costs after the designs have been completed and issued for bidding and construction purposes. Scope changes during the construction phase can be very costly. Finally, there is no guarantee that the lowest bidder will perform adequately.

Figure 8.1 is an organization chart showing the general contractor method of project delivery.

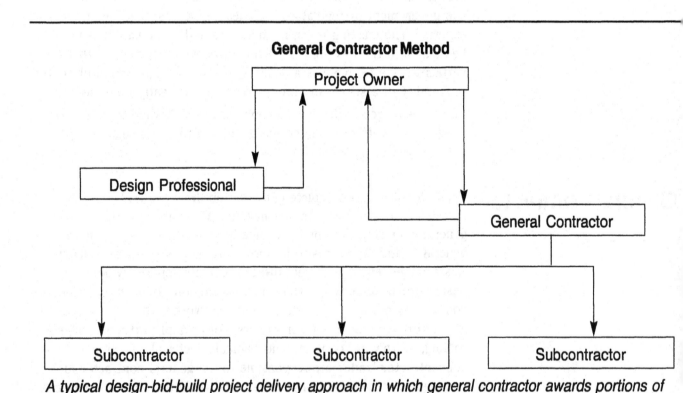

General Contractor Method

A typical design-bid-build project delivery approach in which general contractor awards portions of work to subcontractors

Figure 8.1

182

The general contractor knows that normally he or she must submit the lowest cost bid to the owner in order to get the job. In some cases, this bid may be based on quotes received from less qualified subcontractors. The requirements for a bid bond and a performance payment bond help to establish the financial stability of the contractor and ensure final completion of the project. Although the general contractor design-bid-build method is still widely used, there has been a steady increase in the use of design/build and construction management.

Although the owner-general contractor compensation arrangement may be for a fixed lump sum, owners should still receive unit prices for use in determining equitable compensation for potential additions or deletions to the contract scope of work.

The general contractor is typically selected on the basis of submitting the lowest bid. This can result in problems, as bidders seek the lowest-cost ways of meeting the requirements of the plans and specifications. If the bidding and construction documents are not detailed enough, the low (successful) bidder has the option to make

General Contractor Method

Major Features:
- Two separate contracts one between the owner and design professional, and one between the owner and contractor
- Lump sum competitive bidding or negotiated contract
- General Contractor often selected on basis of lowest bid.

Advantages:
- Lump sum competitive bidding based on detailed and complete construction documents
- Check and balance between architect and contractor
- With high quality, detailed, and well-coordinated construction documents, this method can be an effective project delivery system

Disadvantages:
- History of budget and schedule overruns
- Less owner control over selection of subcontractors (no guarantee lowest bidder will perform adequately)
- Poor General Contractor = Poor Project, contractor alone controls means, methods, & sequence of construction
- General contractor's markup on subcontractors and equipment
- Scope changes during construction are more costly
- Adversarial relationships often develop. Not a team approach to project delivery
- Construction costs determined later in project life cycle than with other delivery methods. Costs cannot be finalized until construction documents are completed
- Design-bid-build approach does not allow phasing of construction (Fast Track), resulting in unnecessarily long completion time
- Subcontractor buyout savings revert to general contractor, not the owner
- Lack of construction input in the design phase, when there is the greatest opportunity to control costs and quality

decisions on materials and workmanship that may not necessarily match the owner's or design professional's assumptions. This is why high quality, detailed and well-coordinated design drawings and specifications are required, along with careful supervision by the project owner as work is being performed.

The general contractor approach can be adversarial because, in a sense, the owner and contractor have opposite financial interests. (Negotiated contracts, however, may include clauses that provide the general contractor with a financial incentive to propose cost-saving ideas that, if accepted by designer and owner, will be advantageous to both owner and general contractor.) The general contractor's potential to realize more profit on a job is in direct proportion to his ability to deliver the work at a lower cost than the bid price. Thus, the general contractor plans, schedules, and controls the project in an effort to minimize his or her costs. The general contractor's relationship with the owner is that of an entrepreneur who has a price to protect rather than a professional advisory or agency relationship. The general contractor is not motivated to share information, such as areas of potential cost savings, that will promote the project owner's interest until he is under contract and it is to his advantage to do so.

The general contractor approach divides responsibility for the project between two separate organizations (design professional and general contractor), with the project owner in the middle. If problems arise, disputes and finger-pointing are frequently the result.

Design/Build

Here, the owner enters into a contract with one firm that takes responsibility for both design and construction of the project. The work may be performed by one company that has both design and construction capabilities, or by a contractor who hires a firm to complete the design work. Design/build contracts are typically negotiated. This method may appeal to owners because preparation of the project budget is included in the design/build service, and therefore there is no outlay of funds unless the project moves forward. The fallacy in this argument is that the estimate is not based on a complete scope of work. Owners must have the technical knowledge and time to understand, monitor, and control the development of the scope of work, in order to ensure that the estimate is reliable and reasonable.

Although most of the responsibility is delegated to an outside party with a design and build contract, the owner is still obligated to perform certain duties such as providing program requirements and design criteria, monitoring progress, paying for work performed, general decision-making, and reviewing and approving plans and specifications. The owner is also responsible for approval of products, making sure that punch list items are addressed and the systems are properly tested and commissioned to ensure conformance with contract requirements. Real estate acquisition, design approval, zoning and permits, accounting and financial control, security, safety, and other responsibilities must be assumed by the owner unless they are specifically delegated to the design/build contractor.

Advantages and Disadvantages of Design/Build

Design/build allows for time savings by overlapping the design and construction phases. There is only one contract to formulate and administer. Contract documents (drawings and specifications) can be simpler due to the partnership between design professional and contractor. There is less impact on your organization's staff and management with only one firm or contact point — a factor that expedites communication for both parties.

On the other hand, there is a lack of checks and balances between contractor and designer. The success of the project depends heavily on the quality of the owner's program documents and performance specifications, and directly on the financial stability, integrity, management controls, and the effectiveness of the design/ build contractor to design, construct, and commission the facility. If the company

Design/Build Method

Major Features:
- Owner enters into a single contract with one firm that takes responsibility for both design and construction

Advantages:
- Time can be saved by overlapping design and construction (fast-track)
- Only one contract to prepare and administer
- Owner works with only one firm or contact point
- Establishes the construction cost early in the design phase of the project life cycle
- Optimal communication between designer and builder

Disadvantages:
- Lack of checks and balances between the contractor and designer
- Success of the project depends on the effectiveness of one company to design, construct, and commission the facility
- Value engineering and life cycle cost analyses (from owner's perspective) are often ignored
- Difficult for the owner to evaluate what he is purchasing without the benefit of detailed scope documents and specifications
- Owner needs technical knowledge and available time or resources to ensure that his organization's interests are protected throughout the project phases

proves to be inadequate in performing any of these duties, it is difficult to remove the contractor without severely affecting the cost and schedule of the project. In addition, contracts must be structured to ensure that the design/build firm does not realize an unreasonable profit. This can be accomplished by requiring the firm to document all costs incurred in connection with completing the contract and requiring the design/build contractor to competitively bid all trade contracts on an "open-book" basis.

Responsibilities must be clearly defined in the design/build contract. The owner should request that the design/build firm provide open-book services, which means that the owner should be able to audit costs throughout the progression of the project to prevent the design build contractor from marking up costs excessively. Properly executed, the design/build approach results in a smooth, efficient, nonadversarial process, and cost-effective design and quality levels that conform with the customer's expectations. Design/build is frequently used for either simple projects, such as commercial office space, or very technical projects, such as chemical or manufacturing process applications. For simple projects, it is not difficult for owner to provide adequate performance criteria. For process-oriented projects, the owner would select a highly specialized design/build fir to help develop the project program and performance requirements. Figure 8.2 is an organization chart showing this method.

Construction Management

During the past 20 years there has been a steady increase in utilizatic of the construction management project delivery method. Project owners are using construction management because it avoids many c the problems associated with the general contractor and design/buil project delivery methods.

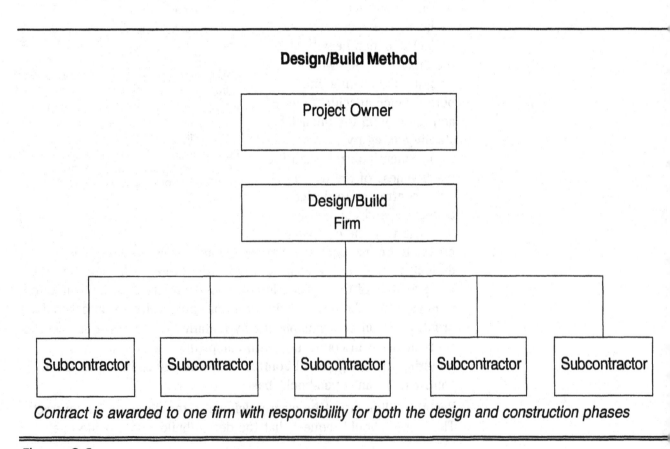

Design/Build Method

Contract is awarded to one firm with responsibility for both the design and construction phases

Figure 8.2

Construction management is a team-based process of organizing and controlling the various independent activities and disciplines that comprise a project in order to best serve the owner's interests. It involves professional business management of the entire building process, encompassing the pre-design, design and construction phases, in order to deliver a project on schedule, at the lowest reasonable cost consistent with the program requirements. Construction management can be performed by an independent professional (agency), a general contractor, or a design professional who provides CM as an extended service over and above design responsibilities. Agency Construction Management is geared toward optimizing the owner's control over cost, time, quality, and risk on projects.

Agency Construction Management

Agency Construction Management is sometimes referred to as "pure" or "professional" construction management. The agency construction manager provides strictly management services in an advisory role as the owner's personal representative throughout the duration of the project. A firm working under an agency construction management contract coordinates activities solely in the owner's best interest. There is a professional relationship between the agency construction manager and the project owner similar to the relationship between an owner and his attorney or accountant.

The potential for conflicts of interest is minimized because the agency construction manager neither provides design services nor performs any construction work with his own forces. All contracts for design services, and all contracts for construction are held by the owner. The result is a good system of checks and balances for the project with the construction manager as manager for the project, the architect or engineer as designer, and the specialty trade contractors performing construction. Each entity can concentrate on what it does best.

The management services provided by a construction management professional can be tailored to the project in the form of a comprehensive contract. The owner can participate in managing the daily project activities as much or as little as he or she desires. The construction manager provides professional business administration of the entire project process. The owner, in effect, manages the project through the construction manager, whose time, cost, and quality motives are the same as those of the owner. The agency construction management approach permits the owner to concentrate his or her time and effort on fulfilling core business responsibilities.

Figure 8.3a is an organization chart showing a typical construction management arrangement.

Contractor Construction Management

In contrast to agency construction management, the contractor construction management firm may agree to perform some of the construction work with its own forces, or subcontract work to specialty trade contractors on an open-book basis with the owner. Since the construction manager is performing actual construction or holding the construction contracts, its role is really no different than that of a general contractor who is selected to provide both preconstruction and construction phase services. Therefore this kind of relationship with the owner is referred to as *contractor construction management.*

Regardless of the type construction management contract selected, the construction manager is most effective when retained at the beginning of the project when the scope is being determined and opportunities for cost savings are greatest. The construction manager can assist the project owner in developing the program requirements and determining project feasibility from a cost, schedule, and constructability standpoint. If hired first, the construction manager

Construction Management Method

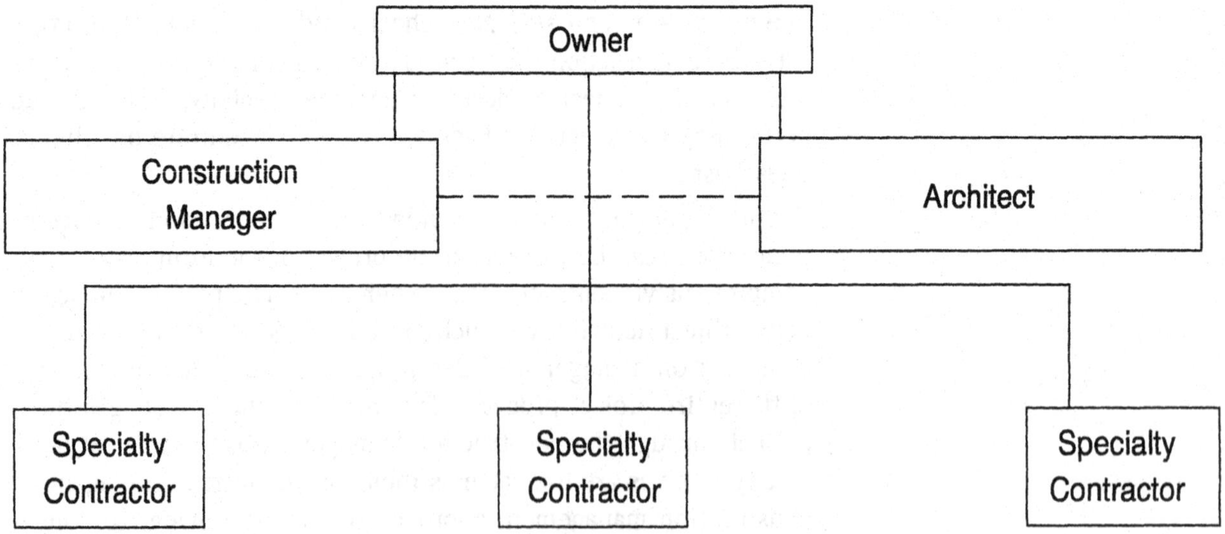

Construction management is a team approach to project delivery. This is often characterized by multiple prime specialty contracts direct with the project owner.

Figure 8.3a

can also help the owner in evaluating and selecting the design professionals.

During the budgeting and early design phase, the construction manager prepares preliminary, or conceptual, cost estimates for alternative design concepts, based on programming and schematic documents. A realistic construction budget is established early in the design phase and should include contingency funds to cover the risk of unknowns within the scope of the project. On construction management projects the owner is privy to more detailed line item budget information than is the case with the general contractor or design/build methods. Construction management budget estimates are typically broken down and classified according to either the Construction Specifications Institute's 16-Division MasterFormat, the Uniformat System, or by contractor work packages. During design development the construction manager monitors and refines the cost estimate, conducts cost-benefit studies, and performs value engineering and related management services to restrain cost creep.

During the planning and preliminary design phase of the project, the construction manager develops a master project schedule which is then monitored and updated until the project is finished. Key decision points and optimum timing cycles for each phase of the project are determined. This is also the time to identify long-lead items and order them early to avoid project delays.

Another advantage of the construction management approach is that it enables the owner to purchase equipment and services directly from the most qualified specialty trade contractors (e.g., earthwork, concrete, mechanical, electrical, masonry, etc.) and equipment manufacturers, rather than having to pay a general contractor who would mark up such items. During the bidding phase, the construction

Construction Management Method

Major Features:
- Team approach to project delivery
- Multiple prime specialty construction contracts direct with the owner or construction manager
- Open competitive bidding
- Central planning and control Construction manager responsible for cost, time, quality control, and risk management (depending on contract arrangement)

Advantages:
- Introduces team approach with checks and balances
- More control by owner
- Allows fast-track
- Cost savings (eliminates general contractor mark-ups on subcontractors and major building systems equipment purchased directly by the owner)
- Construction Manager becomes responsible for effective business management of the project process
- Can adapt quickly to changes
- Establishes construction budget early in the preliminary design phase
- Use of value engineering
- Customized and flexible work arrangements with project owner
- Construction manager applies his specialized knowledge to save money for the owner's budget

Cautions:
- Construction Manager must have the appropriate background and expertise for the particular type or size of project

manager can help prepare front-end bid documents and prepare and negotiate contracts with specialty contractors and equipment suppliers on behalf of the project owner.

The construction manager continues to work with the design and engineering staff by exploring energy, environmental, and operational alternatives. Constructability and maintainability of different design options, and operating and maintenance costs are also considered.

Basic construction services include supervision and coordination of work in the field, quality control, and the preparation of project reports. The construction manager may also provide for various general conditions items and duties that owners may not realize they are responsible for furnishing and performing. These may include owner-furnished materials and equipment; coordination of construction with owner activities; job-site safety; coordination of telephone, data lines, electric, gas, and water utilities; and temporary sanitary facilities. Construction managers also negotiate change orders and review contractors' requests for payments. Additional construction management services include:

- Life cycle costing
- Scheduling
- Identifying and coordinating the purchase of items that have long lead times or items that are to be provided to the contractors by the owner
- Solicitation and analysis of bids
- Contract administration

Another benefit of the construction management approach is that it facilitates the controlled use of "fast-track" or phasing of the construction process, which can result in major time and cost savings. Rather than waiting until an entire building is designed before beginning construction, portions of a project can be contracted out and work can begin as soon as they have been designed and approved. For example, foundation construction can start before the completion of air-conditioning system drawings and specifications. Thus, the design phase can overlap with the construction phase.

The construction manager uses his best efforts to manage the project risks and complete the project in the most expeditious and economical manner consistent with the program requirements. The collaborative efforts of the design professional and the construction manager during the design phase helps to ensure that an economical and efficient design is produced prior to bid, thereby reducing the need for costly changes or redesign during the construction phase. The team approach to management reduces the traditional disputes between the design professional and general contractor and provides the owner with needed continuity and construction expertise from start to finish of the project.

A Word of Warning Many owners, in their haste to lock a contractor construction manager into a guaranteed maximum price (GMP) contract before the construction documents are complete, disregard the expense they incur in so doing. The more detailed and complete the construction documents are, the more accurately the construction manager can estimate the cost of construction. The only defense a construction manager or design/build contractor has against an incomplete set of plans is an increased contingency. A guaranteed maximum price provided early in the design process must contain a substantial contingency allowance to cover the cost of unknowns that would economically favor the contractor construction manager. GMPs provided later in the design process will be more realistic and consequently favor the owner. A GMP with a shared savings clause is usually a conflict of interest because the construction manager will be inclined to price out the project to almost guarantee he will come under the shared savings GMP.

Lump sum or percentage of construction cost contracts invite abuse and therefore are not recommended. The opportunity to compromise quality, whether to make up for cost overruns or to increase revenue, is too tempting. Cost-plus-fixed-fee contracts with a well-defined description of the reimbursable items, and full disclosure of all bids and costs that qualify as reimbursables are recommended.

Figure 8.3b is a sample responsibility matrix for a construction management project. It shows overall and specialized responsibility, as well as a breakdown of the tasks covered by the project participants.

ontract Documents

Once the scope of work and project delivery method have been determined, the project owner must award the appropriate contracts and ensure that the contractors' performance meets contractual requirements. The contract is the basis for allocating risks and responsibilities among the project stakeholders. In devising the contract strategy, owners should first determine which of the risks inherent in project design and construction they are willing and able to assume. A good contract defines the scope of work, provides a set of rules for participants, anticipates potential problems, defines the dispute resolution process, and informs each party of what it must do, by when, as well as what each party is entitled to. The contract is the single most important document the project manager uses to control the project destiny.

Professional societies, institutes, government agencies, and trade organizations have produced standard contract forms that are adaptable to most construction projects. Those published by the Associated General Contractors (AGC), the Engineer's Joint Contract Documents Committee (EJCDC), and the American Institute of Architects (AIA) are commonly used and revised as necessary. Standard contract forms are available for architects' services,

Sample Responsibility Matrix for a Construction Management Project

	General Construction	Owner	Construction Manager	Engineer	Specialty Contractors
1	Project Scope of Work	◎	◆	■	—
2	Conceptual Project Cost Estimate	☐	◆	—	—
3	Project Approach	◎	◇	◆	—
4	Pre-Design Conference	☐	◆	■	—
5	Milestone Schedule	☐	■	◆	—
6	Bid Package Inclusions	☐	■	◆	—
7	Feasibility Study	☐	◆	■	—
8	Design - Critical Date Schedule	☐	■	◆	—
9	General Construction Design	☐	◆	■	—
10	Design Review	◎	◆	◇	—
11	Final Approvals/Design	◎	◆	◇	—
12	Owner Buy-Outs	◎	◇	◆	—
13	Final Project Cost Estimate	☐	■	◆	—
14	Pre-Qualify Bidders	☐	■	—	◇
15	Solicit Bid Interest	☐	■	—	—
16	Assemble Bid Packages	☐	■	◆	—
17	Conduct Pre-Bid Meetings	☐	■	—	—
18	Select Contractors	◎	◇	◆	—
19	Post-Bid Meetings	☐	■	—	
20	Contract Preparation	☐	■	◆	
21	Construction Contracts	◎	◇	—	◇
22	Insurance Requirements	◎	◆	—	◇
23	Project General Requirements	☐	■	—	◇
24	Job Safety	☐	■	—	◇
25	Pre-Construction Meetings	☐	■	—	◇
26	Construction Schedules	☐	◇	—	■
27	Total Project Schedule	☐	■	—	◇
28	Project/Progress Reports	☐	■	—	
29	Construction Cost Control	☐	■	◆	◇
30	Permits (Construction)	☐	■	◆	◇
31	Construction Schedule Update/Assurance	☐	■	—	◇
32	Construction Quality Assurance	☐	■	◆	◇
33	Construction Site Meetings	☐	■	—	◇
34	Independent Consultants	◎	◇	◆	
35	Shop Submittals/Samples	☐	◇	■	◇
36	Progress Payments	◎	◇	—	
37	Scope Changes	◎	◇	◆	
38	Correct Non-Conformance	☐	■	◆	◇
39	Punch List	☐	■	◆	◇
40	Project Close-Out	☐	■	◇	◇
41	Start-Up (Building Systems)	◎	◇	◆	■

Legend: ☐ Overall Responsibility　◎ Overall & Specialized Responsibility　■ Specialized Responsibility　◇ Consulting & Coordination　Consultir

Figure 8.3b

engineers' services, and owner-contractor agreements based on the general contractor, design/build, and construction management project delivery methods. It is always recommended that legal counsel be consulted prior to use of a contract. In addition to the contract agreement, performance bond requirements are sometimes included in construction contracts.

Construction Performance Bond

This bond provides financial protection for the project owner should the contractor fail to complete the work per the agreement. It constitutes a promise by the contractor's surety, or bonding organization, that if the firm does not perform per the terms and conditions of the contract, the surety will take over and complete the work in accordance with the contract.

Construction Payment Bond (Labor & Material Bond)

This type of bond serves as a guarantee by the contractor that subcontractors, material suppliers, and others providing labor, materials, goods, and services to the project will be paid. It gives owners remedies should unpaid parties to the project file mechanic's liens against the owners' property. (A mechanic's lien is a legal instrument which a contractor or subcontractor may file in local court to prevent the owner from gaining clear title to the property until all his or her obligations are met).

Certificates

Various certificates must be furnished by those performing work on the project, indicating compliance with applicable laws and contract terms. They include certificates of insurance, which indicate that parties are insured per the contract, certificates of substantial completion, and occupancy certificates from the building inspector or permit-granting agency.

Drawings and Specifications

The contracts also reference other documents that become part of the agreement, such as drawings and specifications. Drawings graphically depict the work to be done, indicating the type, quantity, and relationships of materials and systems to each other, including sizes, shapes, locations, and connections. Specifications are presented in written form. They are complementary to the drawings and express the quality of workmanship, products, and materials the project owner and designer expects.

Construction specifications are sometimes given minimal attention by the design professionals who are responsible for preparing them, as these professionals may tend to concentrate their efforts on completing the drawings. As the project design period nears completion, a set of specifications are sometimes hurriedly put together. In these cases, the specifications may be incomplete, incorrect, or poorly written; contain contradictory requirements; or show a conflict with information on the drawings. Material is sometimes taken verbatim from earlier specifications that are outdated or inappropriate. These kinds of specifications problems cause confusion, and can possibly result in claims from contractors and consultants for additional compensation.

Recently, there has been a trend for designers to prepare performance-oriented rather than the more traditional procedural or prescriptive-type specifications. The intent is to save time, reduce the probability of errors, and avoid taking on more liability exposure than is absolutely necessary. An example of a performance-oriented specification is prescribing the desired ultimate compressive strength for concrete, rather than giving the contractor a recipe for mixing the concrete in order to achieve the desired strength. Another example is asking the mechanical contractor to supply a product and completed construction based on tons of air-conditioning capacity and cubic feet per minute of air.

Owners, construction managers, and design professionals should be aware of the importance of specifications and allocate sufficient time and money to the preparation of these documents. Drawings and specifications become part of the contract documents when a contract is awarded.

The project manual for a construction contract is typically organized into the bidding documents and contract conditions, and the technical sections (specifications). Bidding documents contain instructions on how the bid is to be prepared and submitted. They apply to prospective bidders interested in the project, whereas contract documents concern only the successful bidders who will be signing the contract for construction with the owners. Bidding requirements typically consist of several documents that are included in the project manual. They cover invitations, instructions and information available to bidders.

Invitation to Bid

This document is designed to attract qualified bidders and help them decide whether to obtain further bidding documents. It includes project identification, description of work, methods of compensation (lump sum or cost reimbursable), and the time and location of the bid opening.

Instructions to Bidders

These include the requirements the firms must comply with before and during submission of their bids. Various requirements include signatures, bonding, inspection of the site prior to bid submittal, information on substitutions, and alternate prices and bid documentation.

Information Available to Bidders

Examples include property survey, material safety data sheets, environmental and geotechnical report data.

Bid Form

A bid form should be included as part of the package that is sent to prospective bidders. The document is prepared by the owner or its consultant and requests data from the bidders. The form contains cost breakdowns, bidders' prices, and alternates and often asks for names of subcontractors and the amounts allotted for specialty construction (e.g., mechanical). The bid form is used to evaluate bids received and to assist in establishing change order prices. Wide variations in line item prices may indicate that the plans and specifications are unclear and subject to interpretation. The bid form can also provide information regarding the suitability of the contractor. By comparing the dollars bid by various contractors, the owner can identify potential gross omissions in the low bidder's proposal. A substantial omission usually leads to an unsatisfactory project, since the contractor may have to create economies wherever possible in order to cover the missing item(s). The owner/project manager can request any amount of detailed cost data on the bid form. Figure 8.4 is a sample bid form.

Bid Bond

The low bidder may be required to furnish a bid bond. This bond obligates a bid-bond surety to pay the owner the amount of the bond, or the difference between the amount of the bid and the next lowest bidder's price should the original low bidder refuse to enter into a contract with the owner. The purpose of the bid bond is to protect the owner from losing the benefit of a low bid. The issuance of the bid bond by a surety implies that the contractor has a satisfactory financial statement to obtain a performance/payment bond.

Conditions of the Contract

In this section, we will examine the two major types of contract conditions: general and supplementary.

General conditions define the rights, responsibilities, and relationships of the parties involved in the construction process. They include general clauses and provisions that are common industry practices. Supplementary conditions modify or supplement the general conditions as necessary to provide for unique requirements. Some inclusions are the time duration of the project, insurance requirements/required wage rates, and liquidated damages clauses. Work restrictions, such as hours of the day when work can be performed, material delivery, storage and staging restrictions, and

Bid Form

1   The undersigned, having familiarized themselves with the local conditions affecting the cost of the work, and with the specifications (including Invitation for Bids, Instructions to Bidders, Bid Form, the form of Contract, and the form of Performance and Payment Bond or Bonds, the General Conditions, the Supplementary Conditions, the General Scope of Work, the Technical Specifications, and the Drawings and Addenda, if any thereto, as prepared by XYZ Design Professionals and on file in their office, 100 Rosewood Lane, Anytown, Anystate 00000, hereby proposes to furnish all labor, materials, equipment, and services required for the construction of a new building for Smith/Jones to be located at 200 Nation's Plaza, Anytown, Anystate 00000, for the following Bid:

Dollars ($ _____ .00)

_____
Written

2   In submitting this bid, it is understood that the right is reserved by the Owner to reject any and all bids. The undersigned agrees to execute and deliver a contract in the prescribed form and furnish the required bond within ten (10) days after the contract is presented for signature.

3. Alternates.

A-1   Cost for adding a termite
      treatment system under the work
      (Spec Section 02281)......... .... Add $17,000 00

A-2   Cost for trench footing systems
      as detailed on drawings in lieu of formed
      footing. (Note: Precast length may be
      shortened by 2") . ... Add $ N.C.

A-3   Cost for adding casework in foyer . ... Add $4,200.00

A-4   Cost of installing the site storm
      drainage system near the east
      property line Include the removal of
      trees and related grading and seeding .... Add $30,000.00

A-5   Cost for the complete underground
      sprinkler system as per plans
      and specifications................. .... Add $20,000 00

A-6   Cost for the complete underground
      lawn sprinkler system (as identified
      in Addendum #1, Item #7, except
      do not include the electric booster
      pump and housing)................ .... Add $26,000 00

Figure 8.4

A-7 Cost for a complete two-part epoxy
floor coating system for the following
areas: 154, 155, 160, & 161. Follow
Specification Section 09726 for basic
requirements or the manufacturer's
recommendation for the following system .... Add $70,000.00

A-8 Provide an alternate roof system to the
one specified, except the roof shall meet the
design requirements of factory mutual 1-90.
This would include a base layer of gypsum
board, the vapor barrier, a single layer of
R-18 extruded polystyrene, a recovery board
and an adhered 60 mil EPDM roof
membrane .... Add $145,000.00

4. Unit Costs

a   Removal and disposal of unacceptable subsoil,
trash, and debris below specified elevations . ... $5 00/cu. yd

b   Provide compacted subsoil fill to replace
excess unacceptable materials stated above . .. . $7.00/cu. yd

c.   Provide one (1) concrete column protector
per Detail Sheet S2.1 . ... $150.00/cu. yd

d   Provide (1) steel bollard per Detail
Sheet S3.1 .... $220 00/cu. yd

5   List of Major Subcontractors·

The lump sum bid noted above includes the name and price quote of the following
prime subcontractor categories:

(The price must be broken down even if one subcontractor has bid several categories.)

a.   The plumbing subcontractor will be Williams-Stover
and their bid amount is $1,162,000.00

b   The HVAC contractor will be _____ Above
and their bid amount is $ _____

c.   The electrical subcontractor will be Metropolitan Tech
and their bid amount is $ 1,172,000 00

d.   The base bid roofing contractor will be Bullock & Sons
and their base bid amount is $360,000.00

gure 8.4 cont.

6. Time of Completion.

The Contractor states that he can complete the work and have substantial completion in
_____ days from the date of contract.

Date: _____ 19 _____

Official Address:

_____

_____

_____

_____

Bidder's Phone No. _____

We acknowledge receipt of the
following addenda:
Addendum #1 dated _____
Addendum #2
Addendum #3

SULLIVAN CONSTRUCTION CORP.
A corporation, incorporated in the State of _____

By: _____

Title: _____

Attest· _____

Figure 8.4 cont.

security requirements, are factors that affect productivity and should be included as part of the supplementary conditions.

When liquidated damages are prescribed, the contractor agrees to pay the owner an agreed-upon amount of money if the owner cannot gain beneficial occupancy of the project by the time specified in the contract. Retainages may also be included. Where liquidated damages are prescribed, many legal jurisdictions require that the contractors receive bonus money from project owners for each day ahead of the specified contract completion deadline that they are able to complete the project.

A retainage is a stipulated amount of money withheld from payment by the owner until project completion to ensure that the contractor completes the specified work. It is customary to retain 10% of payments due until the project is 50% complete, then reduce the amount of retainage from 10% to 5% after the project is 50% complete and until contract performance is ensured.

Potential Owner Liabilities

The following list includes situations in which potential liabilities could be the owner's responsibility:

- Safety
- Project delays or disruptions: Examples include late delivery of owner-furnished items, such as construction drawings or owner-furnished equipment; untimely approval of change order requests, shop drawings or samples; deficient construction documents; or failure to provide timely inspection of completed work.
- Site availability problems.
- Interference with contractors' schedules and ordering them to proceed under adverse conditions.
- Failure to properly and promptly perform certain tasks which precede the contractor's work.
- Erroneous information that misleads or disrupts the contractor.
- Failure to disclose information necessary for satisfactory contractor performance.
- Directing contractor's operation.
- Requiring contractor to use a particular method when the contract does not specify any particular method. Note: If the owner requires a specific methodology, it must be specified in the contract documents. Specified methods usually relate to safety requirements (beyond OSHA) and staging of work to comply with owner's work/production schedule.
- Failure to provide inspectors for field tests conducted in accordance with contract documents.
- Overinspection: In some cases, owners may require installation and placement tolerances of materials and equipment that are tighter than customarily required. The tolerances/requirements

must be stated in the specifications. (If a higher level of performance than is specified is imposed, the contractor may be entitled to an equitable adjustment in the contract sum.)

- Increased owner requirements causing unnecessary interference with contractor's work, such as test requirements beyond those listed in the specifications.
- Failure to make timely payments.

Since these are only some of the potential risks, the owner may want to assign responsibility for managing project risks to a professional construction manager (CM).

The owner also should be aware of the necessity of job-site safety programs and adequate insurance protection.

Job Site Safety Programs

Safety is everyone's concern. The doctrine of comparative negligence is used by courts of law to weigh the relative negligence of parties involved in a claim for compensation arising from injury due to unsafe site conditions. After weighing the relative negligence, the courts allocate a percentage of responsibility to the parties involved. In many cases, "negligence" is determined by the plaintiff's attorney. The case does not go to court. The settlement is based on the amount of insurance coverage available.

Owners should take active roles in enhancing safety throughout the project by developing their own job site programs. A list of addresses and telephone numbers of doctors, hospitals, and ambulance services to be used in emergencies should be compiled and posted conspicuously at the job site and attached to telephones. Safety awareness signs should also be posted as determined by law, along with a fire extinguisher and an OSHA-approved first aid kit.

Owner Responsibilities

Furnish detailed information regarding the project requirements (scope of work), including·

A Detailed survey of property, including elevations, topography, utilities, and building setbacks

B Budget for construction costs

C Equipment list and process requirements.

Also

- Review proposed preliminary drawings
- Sign off on final preliminary drawings and budget price
- Approve contracts for bidding purposes
- Approve successful bids

Conducting regular tours of the job site helps to ensure compliance. If an unsafe act or condition is noticed, it should be brought to the attention of the contractor's superintendent for prompt corrective action. It also should be noted in the owner's project log.

Letters and memorandums to the contractor's superintendent are not advised. Safety is the contractor's responsibility, and a written letter from the owner to the contractor may imply that the owner is assuming part of the contractor's responsibility. However, if an unsafe condition is not corrected, then it may be necessary for the owner to issue a written order to either correct it or suspend work until it is corrected.

Power lines and utilities may also present safety hazards. It is important that overhead and underground power lines be deactivated in the construction zone. Utilities should be notified prior to excavation and construction so that they can locate and deactivate lines that may create unsafe conditions.

Finally, owners should help guard against potential accidents due to collisions involving their own equipment, such as forklift trucks; or their production operations. Basically, project owners must be able to handle situations in which clear responsibility may not be attributed to individual contractors.

Insurance Protection

An independent insurance professional should review a contract before it is signed to make sure that the owner is adequately protected before a construction project gets under way. Contracts should be scrutinized to determine the actual property or persons protected and the events covered by the insurance. The owner should also purchase Builders' Risk Property and Owner Liability insurance to protect against possible claims as a result of the construction project. The owner thus will have direct control over what property is insured and what claims are protected against.

The owner's choice of insurance company should be based on an insurer's record as an adjuster of losses. The insurance company chosen may or may not be the same one that provides other policies needed during the normal course of business operation.

The owner should also obtain certificates of insurance coverage from project participants prior to their starting work. The owner should be listed as an additional named insured party. Consultants are often included as additional insureds on the insurance certificate. Architects, engineers, surveyors, and other professionals on the project should carry liability insurance to cover errors or omissions arising from negligent performance and failure to perform professional services. "Errors and Omissions" coverage, carried by design professions, is the principal source of funds from which judgments can be recovered.

Construction contractors should carry Broad Form comprehensive insurance to cover liability arising out of the use of premises at the project site. Contractors assuming design or CM responsibilities should also carry liability insurance.

In general, contractors should purchase and maintain insurance to protect against claims such as the following:

1. Workers' Compensation, disability, and similar benefits established by state statutory requirements.

2. Claims for damages because of bodily injury, or occupation-related sickness, or disease or deaths of employees.
3. Claims for damages because of bodily injury or occupation-related sickness, or disease or deaths of persons other than the contractor's employees.
4. Claims for damages for personal liability as a result of an offens directly or indirectly related to employment of a person by the contractor or any other person.
5. Damages other than to the work itself resulting from damage to or destruction of tangible property.
6. Claims for damages because of bodily injury or death of any person, or property damage arising out of ownership or use of a vehicle.

Methods of Compensation

There are numerous fee structures for construction projects. Use depends on factors such as degree of uncertainty faced and competitive environment. In every case, however, the owner's objective is to assume the least amount of risk possible while still offering enough incentive to the consultant and contractor to ensure delivery of a cost-effective, high-quality product. At the same time, contractors and consultants naturally seek to reduce their own risks while maximizing net gains. Thus, the fee structure that is ultimately chosen must be satisfactory to both parties to the contract.

There are two general categories of fee structures: *lump sum* (fixed price) and *cost reimbursable*. There also are some variations: *fixed pric plus incentive*, *cost plus incentive*, *cost plus fixed fee*, *cost plus percent of cost*, and *unit price*. Performance-based contracts based on a percentage of cost savings are another compensation method. Each type has advantages and disadvantages.

Lump Sum

In the lump sum fee structure, a firm price is established before awarding the contract. The price presumably remains fixed and guaranteed for the duration of the contract and covers all work specified in the construction documents. The lump sum covers the contractor's direct costs (labor, material and equipment), indirect costs (home-office supervision, administration and secretarial support), and profit margin. Detailed construction drawings and specifications must be available for contractors to guarantee performance of the work for a fixed price.

This fee structure enables the owner to know exactly how much to budget. The amount will not change, except for authorized change orders. The owner's risk is minimized. However, a contractor who has not budgeted enough money to complete the project may cut corners at the owner's expense, substituting lower quality material c products, providing inferior workmanship, or decreasing supervisio of the project — thus failing to complete it in accordance with the established objectives. If there is a poorly defined scope of work and inferior contract documents, costly design and construction changes may result and add to the lump sum price.

Fixed Price Plus Incentive

This guaranteed-maximum fee structure is a slight modification of the lump sum type. It allows for either increases or decreases in the basic fee, depending on the disposition of certain factors and costs incurred by the contractor. The contract that accompanies this type of fee structure is very complex. Five key factors must be included: *target cost, target profit, target price, ceiling price,* and *share ratio.*

The fixed price plus incentive fee structure provides a profit incentive for the contractor to reduce costs and maximize efficiency. Risks and advantages are balanced between owner and contractor. However, after the contract is signed, the owner who discovers that the ceiling price or guaranteed maximum is set inordinately high has no recourse. If the scope of work is indefinite or unclear, the owner and contractor may agree to a higher ceiling price to help absorb the costs of probable changes.

Cost Plus Incentive

In the cost plus incentive fee structure, the contractor is reimbursed for all costs and is offered an incentive for keeping them lower than the target price. If the final cost is lower than expected, both owner and contractor benefit according to a pre-negotiated formula that applies a percentage ratio to adjust fees. Let's say a project has a target price of $200,000, with the contractor's fee set at $20,000. The agreed-upon fee adjustment formula is an 80/20 percentage ratio, in which the contractor absorbs 20% of the risk (and reward) and the owner receives the balance. If final costs are determined to be only $150,000 (a $50,000 savings), the contractor will receive $150,000 actual final costs, including the agreed-upon $20,000 fee, and a $10,000 incentive payment (20% of the $50,000 savings). The owner realizes a $40,000 reduction in the target price — 80% of the total $50,000 savings.

Cost plus incentive offers advantages to both owner and contractor if they agree to share risks. There is a strong incentive for the contractor to control costs without compromising quality. This can be attractive where the scope of work has been well-defined at the time the contract is negotiated. The incentive feature tends to reduce the costs of subsequent mid-project changes.

Cost Plus Fixed Fee

In this arrangement, the contractor is reimbursed for all costs and receives a fixed fee that does not change unless the scope of work changes. Generally, this fee is calculated using a percentage based on the cost, duration, and difficulty of the task. Consultants are often paid under this type of fee structure when the complete scope of work is not known. This fee structure is commonly used when it is difficult to estimate the amount of work and time requirements on projects such as a complex remodeling job.

Cost Plus Percent of Cost

This fee structure requires the owner to reimburse the contractor for all costs, plus pay an additional percentage of these costs as a fee. This is the owner's most risky arrangement. The contractor has little incentive to control costs and is, in fact, rewarded more as they escalate. The owner can exercise some control with a contract that allows for adjustments to the fee percentage rate during the project, if costs exceed original estimates.

This fee structure is selected when the scope of work is not known, and detailed drawings and specifications are not available. Used well, this arrangement can be quite flexible. If this fee structure is selected, the owner must be prudent in selecting a contractor. The owner also must plan to spend a lot of time scrutinizing and monitoring contractor costs once the project begins.

Unit Price

The unit price method of compensation is used when the extent of work or actual quantities cannot be fully determined at the time the contract is signed. Examples include earthwork, where the suitability of subsoil conditions cannot be easily foreseen. Bidders will not risk contracting to perform work for a fixed or lump sum rate if the amount of work required is not known.

Earthwork is often performed on a unit price basis. No matter how many soil tests and borings are taken, it is often not possible to determine the extent of work to be completed. Bids are taken on the cost per unit of measurement. The earthwork contractor will charge the owner on the basis of cubic yards of earth excavated or engineered fill material used. The contractor estimates the quantity and submits a bid for the contract based on a unit price. As the site work is performed, actual quantities are measured and the contractor is paid according to the quoted unit prices. Square feet of drywall, lineal feet of conduit, lineal feet of trench, and cubic yards of concrete are other examples of construction elements that can be broken down into estimated quantities in terms of unit prices. Unit costs are also helpful in ensuring that the project owner gets fair pricing when the scope of work for a particular trade is expanded, or a fair reduction in the contract price when the scope of work is reduced.

Creative Ways to Finance and Contract New Projects

Due to budget constraints, many facility managers are under constant pressure to find innovative ways to increase operational efficiency without a capital outlay. Performance contracting can provide a way for owners and managers to obtain needed financing for projects that improve facilities' efficiency, productivity, and profitability.

In some cases, equipment manufacturers and contractors are willing to upgrade or replace older, less efficient HVAC, lighting and power systems with more reliable, energy-efficient equipment at no cost — in return for a percentage of the savings. This type of arrangement is often called *performance contracting*. A performance contract can be used for any project that directly affects savings or profitability. However, it is more commonly used to finance energy conservation projects. Certain energy utilities, engineering firms, mechanical and electrical contractors, property management firms, and consultants are providers of performance contracting services. There are also third-party agents who are interested in financing energy-related performance contracts. Some outsource companies who work on a performance contract basis will procure the fuel, and operate and maintain the facility infrastructure.

On the surface, performance contracting appears to be a win/win/win situation for the equipment manufacturer, contractor, and building owner. The owner avoids an up-front capital outlay, while lowering day-to-day operating expenses. The contractor and equipment vendor who provide the funds, equipment, and labor for the improvement project enjoy revenues over a predetermined period of time based on a portion of the savings. However, there must be detailed guidelines for measuring and documenting the performance of systems installed under performance contracts. It may be costly and difficult for all parties to a performance contract to substantiate and quantify the savings over time.

Summary

No single contracting method or fee arrangement is best under all circumstances. Facility managers need to be aware of the alternatives and be able to work with their legal advisors to put together the most appropriate contract for a given project. The next chapter explains how to protect the owner's interests in another way — by creating and sustaining good communications to promote the teamwork essential to successful projects.

Part III

Improved Productivity Through Communications and Teamwork

Chapter 9

Good Communication: Vital to Productivity Improvement

Communication is a basic skill that is needed to establish and maintain productive relationships. A high percentage of the friction, confusion, frustration, disputes, and inefficiencies in our working relationships are traceable to poor communication. Facility management work is especially susceptible to communication problems because of the broad and overlapping areas of responsibility, and the multidisciplinary and sometimes complex nature of facility-related projects. Facility managers and facility engineers must be aware of the importance of communication in a building services environment to ensure people are working together effectively to meet organizational objectives. And, of course, facility managers need to develop better oral and written communication skills themselves.

Communication problems are an enormous threat to profitability. In almost every case, the misinterpretation of a customer's requirements, failure to execute a plan or carry out instructions, or a missed delivery date is a result of a breakdown in communications. Applying some fundamental principles and techniques to this area will lead to more effective use of available time, improved cooperation and coordination of efforts, fewer disputes, and a reduction in associated costs.

Why do communication breakdowns exist in our Age of Information Technology? Messages and images can be transmitted instantaneously almost anywhere in the world. Through use of the Internet combined with high-speed computers and reproduction equipment, we can distribute reports and other information, in real time, almost anywhere at any time. Communication is a process. The key is that you have to receive and transmit the *right* information on a *timely* basis in order to maintain control. The fax and the computer sometimes overload us with information, and we have to make judgments about what is truly useful Effective communication, versus mere sending of data, means the person receiving the message must *understand* the message and be *motivated* to take the action recommended by the sender. The cycle must be completed for communication to be successful. Transmitting information is the easy part. The problem is reception — more precisely, intelligent reception — of the information.

Also, the receiver must be able to secure clarification and additional information. The sender, in turn, must have feedback enabling him to assess the degree of understanding and compliance. Through feedback, the sender determines requirements for new or follow-up communication. Successful communication, therefore, lies in making a two-way process: downward (from sender to receiver) and upward (from receiver to sender). Distributing established schedules and status reports are examples of one-way communication, while project team planning meetings, status review meetings, value engineering workshops, and post-project review meetings are examples of interactive two-way communication.

The only way to have feedback is to include an aspect of communication often overlooked — listening. Communication cannot be two-way unless we listen actively and with sensitivity. In the best working environments, people feel free to express their views, knowing that their opinions will be considered, and their suggestions and ideas recognized.

There is an old maxim: "If you talk too much, you can't hear what others are saying!" Experience demonstrates that by listening long enough, one starts to get answers. One hears others define the problems and suggest answers. This is considerably more helpful than evaluating only what one person thinks from a vantage point that may be far removed from the problem. It is also important to try to understand the other person's point of view.

Listening can be disturbing because it sometimes forces you to recognize unexpected problems. Often it is more comfortable not to listen, and to ignore warning signs that will require involvement in solving a problem. Another obstacle to effective listening is ego, which prevents one from hearing another point of view. For example an architect's mission is not to design a "trophy" building for his or her own accolades, but rather to fulfill the customer's needs. The grand entrance or spiral staircase may not always be cost-justifiable. The focus should not be on "me," but on the customer's perspective.

Often people will withhold communication of information (such as a potential problem) if they fear somebody will not respond favorably it. Remember the old maxim, "Don't kill the messenger." The fact is that major problems can often be avoided if warning signs are recognized and reported while there are still opportunities for corrective actions. Many problems can be minimized or avoided if somebody has the courage to communicate bad news. Facility managers should instruct their staff to listen carefully, be succinct, and follow through by taking the appropriate action after communicating with the right individuals in a timely fashion.

Project Communications

Early warning signs that a project could be heading for trouble include:

- Delays (schedule slippage, request for time extensions)
- Breakdown in communications (Common causes include incomplete, inaccurate, or untimely transmittal of shop drawings, and cost and schedule reports.)
- Slow payments to consultants, equipment vendors, and contractors.
- Substantial increase in change order requests and claims
- Inefficient crew sizes
- Complaints from consultants, contractors, and vendors
- Quality defects
- Abnormal number of contractor requests for substitutions
- Increasing number of contractor requests for information
- Deteriorating supervision

Part of good communication is learning your customers' culture and what they need. Be a counselor to your customers and support them constantly throughout your working relationship. Emphasize the two-way communication process: They need to be informed, and you need to request their feedback constantly in the form of suggestions and advice. You can keep them informed through vehicles such as regularly scheduled briefing sessions and routine reports and written plans.

At times, facility managers may get involved with a project only to discover that team members are reluctant to share information that may be crucial to successful completion. This is a normal protective human trait. Managers who sensitively convince their staff that they are not trying to assign blame will be more apt to obtain information about conditions that may impede progress. Perhaps a safety hazard needs to be removed or more efficient equipment installed. When people are preoccupied with personal defensiveness rather than organizational objectives, they are not working together effectively to complete their assignments.

To gain cooperation, managers should share their own experiences with similar situations to create an open environment where people can freely share their ideas. Staff members, consultants, or contractors who are continually criticized for sharing their ideas will stop sharing them. Despite your time constraints, it is important to demonstrate consideration for new ideas. Sometimes what appears at first glance to be the most ridiculous idea, turns out to be the best one.

Competent executives, managers, and staff specialists often diminish their effectiveness by maintaining only one-way interpersonal contact. They issue hurried or unclear instructions. They initiate change orders, but fail to ask whether important points are grasped. They assume that others use and understand precisely the same terminology, and often ignore suggestions. With one-way directives, they are actually talking without listening. Managers must take time to communicate thoughtfully — both orally and in writing.

"Communication" is a term seldom found written into contract provisions. Nevertheless, you can write better contracts with a complete statement of the work to be done, and a definition of the roles and relationships between the parties to the contract, lines of communication, frequency of team meetings, and relevant reporting and report formats. Everyone involved in a project should be provided with a project directory containing a list of important telephone numbers. The project manager should have the home as well as the work and pager or cell phone numbers for key project players. The home phone number is important to permit quick answers or exchange of information that cannot wait until the next day.

Written Communications

Communication is at its best when it is clear and to the point, particularly when customers are not experienced or familiar with industry jargon. Moreover, complicated quantitative data is often more understandable to readers when presented in graphic form. Much of the material included in reports is more easily understood when the written information is accompanied by graphic aids such as tables, bar charts, pie charts, and graphs.

Important messages delivered orally must be followed up in writing. Meeting minutes and memoranda must be given to project team members to ensure documentation of important information. Keep project team members informed by regularly sending copies of correspondence, progress reports, calculations, and other significant project documentation.

Shop Drawings

As mentioned in Chapter 8, construction drawings and specifications communicate in graphic and written form the designers' expectations for the work to be performed by the contractor(s). The construction documents prepared by the design professionals do not show all details that affect the constructed product to be put in place. Also, design professionals typically do not communicate means, methods, sequence of construction, or related job safety procedures which are typically the contractors' responsibility. The contract between the owner and contractor(s) should state that the contractor(s) submit more detailed drawings, diagrams, and schedules for review and approval by the design professional (architect and/or engineer) prio to beginning construction. These shop drawings or submittals serve as important feedback from the construction or installation contractors, signifying that they understand the design intent and the contract documents.

Meetings, Meetings, Meetings

Meetings are required to complete — on-time and on-budget — projects that require the involvement of a multi-disciplined team. Meetings offer a sure and fast way to share information among a group of individuals because they provide instant two-way communication. Every participant obtains an immediate response to a question, or clarification of unclear points. Properly run, meetings also save considerable time that otherwise would be spent sending memos, waiting for responses to the memos, or answering letters.

Successful meetings must be planned. Many meetings are too long or attempt to accomplish too much. Meetings should be conducted only when necessary and planned with a clear objective in mind. To properly plan, conduct and control productive meetings, you need to define the objectives and prepare an agenda. Preparing an agenda not only provides you with a tool to control the meeting, but also demonstrates that you have done your research and are prepared to make the most productive use of other people's time.

Project orientation meetings enable management to review the project requirements and obtain the cooperation of consultants and contractors. Project orientation and "kick-off" sessions should be designed to establish a "we" attitude, rather than the "me" outlook that might be prevalent with the involvement of numerous independent consultants and contractors. A "we" attitude transforms a group of people working on a job into a working team doing a job. Regular meetings for clients and contractors, for instance, could include a complete, but brief, report of job progress. Decision-making meetings may solve specific problems such as a productivity issue, pointing out the required corrective action.

Project Record Filing System

Documentation is an essential aspect of communication. Records of design criteria and scope of work serve as a basis for identifying, documenting, and reporting changes. A record file containing sketches, charts, telephone conversations, notes, proposals, estimates, schedules, and other important communications should be maintained for a complete record of the project's evolution. Each item should be dated, and minutes included so that meetings and agreements can be recorded. Not only does careful documentation protect your interests for the current project, but it also provides a storehouse of knowledge for application to future projects.

A structured project record filing system should be maintained in a central location. Duplicates of contracts should be kept in several locations for protection in the event of fire or other losses and as a defense against claims, project overruns, and delays. Incoming and outgoing correspondence should be differentiated. Responsibility for the maintenance and security of files should be delegated to one person who can help prevent their disappearance by controlling file removal and access. Figure 9.1 is a sample procedural outline for a project record filing system.

- Project Description
- Budget
- Equipment List
- Contracts
- Meeting Minutes and Telephone Memorandums
- Correspondence and Transmittals
- Engineering Information
- Estimates
- Specifications
- Request for Bid Packages
- Bid Analysis Documents
- Contractor Proposals
- Insurance Certificates
- Submittals
- Schedules
- Permits
- Reports
- Change Orders
- Utility Agreements
- Notes

Photographs provide excellent records and should be taken (and dated) of site conditions, including the conditions of nearby building and site characteristics, prior to the start of construction projects. Foundations, wall systems, parking lots, and roads should be included. Photographs serve as important records in disputes.

A transmittal letter (see Figure 9.2) should accompany all important documents and submittals so that the sender has a record of what was sent. Owners should ensure that minutes are kept, and validate their understanding of what occurred at meetings based on the recorded minutes. If owners receive minutes with which they disagree, they should issue written clarifications. Minutes are also used to communicate the team's understanding of the project requirements.

Following these guidelines for an organized documentation system will improve productivity, as it saves project personnel from wasting time trying to locate reference material or duplicate previous efforts.

Procedures for Project Record Filing

ob information sheet is filled out by the Project Manager and given to the Accounting Department, listing.

A. Client
B. Job Name
C. Location
D. Client Contact

E. Type of Contract
F. Contract Price if Applicable
G. Estimate if Applicable
H. Large or Small Job

e Accounting Department will complete the Contract Data Sheet, issue a job number, fill out an estimate/ntract sheet, and enter necessary information to the "Jobs in Process" binder.

counting Department will collect all new jobs and enter them into the cost system the following week

es are prepared as follows:

b Files contain:

1. Green pendaflex indicating job number and identity.
2. Manila folders will be made up with the following tabs:
   a. Contracts and Approved Change Orders
   b. Quotations, Estimates and Proposals
   c. Correspondence
   d. Engineering
   e. Purchase Orders
   f. Job Information
   g. A separate file is made up for the Accounts Payable/Billing in Process drawer for bookkeeping and billing procedures.
   h. Invoices
3. Extra work order (a field-generated change in contract amount due to unanticipated work).
   a. Rough draft is prepared by Project Manager and given to Accounting Department listing the following:
      1. Description of work
      2. Estimated hours worked on job
      3. Materials
      4. Equipment
   b. The Accounting Department will compile labor dollars and extend figures into final cost
   c. Finished rough draft is reviewed with Project Manager and, when approved, is initialed.
   d. Extra work is then prepared in final form.
   e. Contract data sheet is updated by Accounting.
4. Change Order—An office-generated change order in contract amount due to change in job. (Processing is the same as noted above for the extra work order.)
5. Sections providing information on the overall project generally contain the following files:
   a. Specifications

gure 9.1

215

b. Architects/Engineers

c. Contractors

d. Contracts/Approved Change Orders

e. Correspondence

f. Transmittals ("Incoming" and "Outgoing" arranged chronologically)

g. Design Criteria

h. Estimates

i. Field Notes

j. General

k. Insurance

l. Job Information

m. Memoranda

n. Meeting Minutes, Telephone Logs, Field Observation Reports

o. Permits

p. Project Criteria

q. Purchase Orders

r. Change Orders

s. Schedules

t. Pertinent Documentation

u. Shop Drawings

v. Transmittal Copies (Alphabetical)

w. Vendor Quotations and Proposal Evaluations

x. Zoning

y. Contract Close-out Documents

6 For large projects, files may be set up by contractor work packages or classified by CSI MasterFormat Divisions:

Div. 1 - General Requirements

Div. 2 - Site Work

Div. 3 - Concrete

Div. 4 - Masonry

Div. 5 - Metals

Div. 6 - Wood and Plastics

Div. 7 - Thermal and Moisture Protection

Div. 8 - Doors and Windows

Div 9 - Finishes

Div. 10 - Specialties

Div. 11 - Equipment

Div. 12 - Furnishings

Div. 13 - Special Construction

Div. 14 - Conveying Systems

Div. 15 - Mechanical

Div 16 - Electrical

Figure 9.1 cont.

ETTER
F TRANSMITTAL

OM

_____	DATE _____
_____	PROJECT _____
_____	LOCATION _____
_____	ATTENTION _____
_____	RE _____
_____	_____
_____	_____
_____	_____
_____	_____

entlemen:

E ARE SENDING YOU ☐ HEREWITH ☐ DELIVERED BY HAND ☐ UNDER SEPARATE COVER
A _____ THE FOLLOWING ITEMS·

☐ PLANS ☐ PRINTS ☐ SHOP DRAWINGS ☐ SAMPLES ☐ SPECIFICATIONS
☐ ESTIMATES ☐ COPY OF LETTER ☐ _____

COPIES	DATE OR NO	DESCRIPTION

HESE ARE TRANSMITTED AS INDICATED BELOW

☐ FOR YOUR USE   ☐ APPROVED AS NOTED   ☐ RETURN _____ CORRECTED PRINTS
☐ FOR APPROVAL   ☐ APPROVED FOR CONSTRUCTION ☐ SUBMIT _____ COPIES FOR _____
☐ AS REQUESTED   ☐ RETURNED FOR CORRECTIONS   ☐ RESUBMIT_____ COPIES FOR _____
☐ FOR REVIEW AND COMMENT ☐ RETURNED AFTER LOAN TO US ☐ FOR BIDS DUE _____
☐ _____

EMARKS _____
_____
_____
_____
_____
_____

ENCLOSURES ARE NOT AS INDICATED,
LEASE NOTIFY US AT ONCE   SIGNED _____

Figure 9.2

Handling Change Orders

Change orders that occur during the life cycle of a construction project also demand timely and accurate communication. They represent a change in the cost and time, from that originally planned and budgeted. Depending on the scope of the change, these situations can be of considerable concern to a project owner.

When changes occur, the project manager must move quickly to provide the customer with adequate documentation concerning the cause of the change. Changes can be the result of a design omission, an unexpected occurrence, or a change in the customer's preference. Cost estimates should be provided in a timely fashion, and the necessary work to implement the change should be scheduled. Regular progress meetings and update reports will help prevent time and budget surprises, and will enhance the owner-project manager relationship.

Time is of the essence not only for change orders, but for other project communications. With fax, email, and voice mail systems readily available, customers not only expect, but demand, timely communications. Reporting must be prompt. They identify potential problems, thereby making it possible to take corrective measures to prevent small problems from turning into big ones. Modern project management software systems allow for relatively easy updates. Variance analysis can also be used to identify problems and to determine the reasons for the variance.

On-Site Communication

The owner's site representative or contractor's field superintendent plays a special communication role. This is a leadership position, that goes far beyond policing contractors to make sure they are working and that construction materials are delivered on time. The owner's site representative's most important task is to serve as the company liaison with the "workers in the trenches." A variety of craftsmen skillfully make the designer's drawings and specifications come to life with their building skills. The site representative has to deal with a wide range of personalities and situations daily. Each worker comes the job with his own unique worries, long-standing attitudes, prejudices, possible health and other issues. Each person working on the project draws conclusions, expresses emotions, and is often influenced by inaccurate reports from the "grapevine." It is important to work with the whole individual, not just that person's engineering, carpentry, or administrative competencies.

Regardless of whether the worker is a corporate employee, an on-site laborer or a consultant, that individual's freedom to contribute opinions on the project allows him or her to feel involved and important. Whether the work is new construction or ongoing maintenance, managers benefit from the suggestions of workers closest to the project. Managers must learn to hear the unspoken, or that which may not be explicitly stated. For example, when asked,

"How many times will we have to repair this broken-down equipment?" the response should not be merely a number. The manager should ask if they are trying to say that the equipment has become worn to a point that it is interfering with productivity. It is important to listen for the meaning behind certain communications.

A Word About Advanced Communications Technology

Owners and facility managers can utilize proven new communication technologies to reengineer facility management practices. Communicating via fax and e-mail reduces the time period for delivering a project. Collaborative work technologies are producing major changes in the way construction projects are procured and facilities are designed, built, and managed.

In the past, long distance business relationships were relatively scarce and costly. It was not productive for designers and engineers, who were geographically separated, to collaborate on projects. Much time was wasted waiting for information that was sent via U.S. mail (snail mail), and it was difficult to process or decipher information sent on a piecemeal basis. Today, communications technologies enable organizations that may be geographically remote to work together on projects simultaneously.

Video conferencing enables face-to-face teamwork over distances. Internet access and groupware programs provide instantaneous electronic access to complex information from remote locations. Portions of a project can be assigned to businesses located across the globe in order to work on a project 24 hours a day. The services of lower-cost design and technical consultants from any locale can be retained to minimize design costs without compromising quality and performance. When consultants are required to travel by air, it is much more convenient and affordable than in prior years.

Capitalizing on developments in communication technologies provides a way to get more use out of less space. It is now a generally accepted practice for sales and consulting personnel who frequently work outside the main office to telecommute. By working primarily out of their homes, they reduce office space needs and the associated cost. When necessary they can reserve space at the main office. Facility managers who need to avoid unnecessary costs can take advantage of these same arrangements.

Computer-Aided Facility Management (CAFM) programs can be used to store, organize, and process large amounts of facilities data in a variety of ways. Computerizing facility management functions enables companies to systematize their operations, handle more inquiries, and establish uniform standards for maximum efficiency.

Computer-aided building management systems can be installed to monitor and control the condition of systems and equipment from remote locations. For example, in 1987, the resident operating engineer was required to be on the premises to monitor operation of the physical plant. Now the computer can perform many of these functions, and it will summon the mechanic when needed. When equipment is operating outside a control set point, the system automatically contacts the mechanic and other designated personne via beeper or a communication alternative. Remote adjustments can be made through a laptop or other computer. Automating facility management functions can simultaneously improve performance and reduce overhead. Fewer personnel are required to operate and maintain an automated facility than are needed for a plant that is not automated.

The merging of computer and communication technology enables facility managers to maximize productivity of the work environment. This increase in automation is costly at first. It would be unreasonable to expect an immediate payback from an investment in automation, as the new systems must operate alongside the old ones for a period of time. However, the long-run benefits from productivity improvement may far outweigh the initial cost. Careful analysis is required before investing substantial funds in automating a facility management operation.

Summary

Productive work environments require good upward and downward communication, clear channels of communication, and successful adoption of new computer and telecommunications technology. Effective communication is the backbone of a productive, team-based work environment.

Chapter 10

Team Problem-Solving and Decision-Making Techniques

Communication clearly plays an important role in team building and facility management. Make a rule of encouraging team members to participate in the planning and implementation of work or events that will affect them. There are three benefits to this approach: you may get some valuable ideas, team members will better understand the reasons for the decisions and actions taken, and they will see you are sensitive to their needs and motivations.

Assembling of Contractors and Consultants for the Project Team

The way in which the team is formed is important if you want to optimize a team approach throughout a project. Consultants, contractors, and suppliers should be carefully selected and managed under the terms and conditions of well-defined contracts. When you retain these services, you are engaging in a partnership. You and they are partners working together to accomplish the common goals of the project.

Select the best qualified consultants, rather than emphasizing only low cost. By recognizing the value of qualified designers, engineers, construction managers, contractors, and suppliers, the owner reinforces the team approach which serves the project's objectives.

Selection of consultants or contractors should be based on professional and technical qualifications. You may want to consider forming a committee to help you develop qualification requirements, and to prepare written requests for proposal forms. If there is any question about why the consultant was selected, you will have a basis for justifying your decision.

Owners should prequalify contractors and consultants, and require them to fill out standard questionnaires that request technical and financial information. References from their previous projects are invaluable. AIA Document A305, "Contractor's Qualification Statement," is a good form to pre-qualify bidders. It can be obtained from The American Institute of Architects' headquarters in

Washington, D.C. Careful evaluation of consultants and construction firms prior to entering into a contract is a major factor in ensuring quality and performance, and can determine the owner's ability to obtain the best facility for the money.

In evaluating consultants, contractors, and suppliers, keep in mind that it is time-consuming and expensive for them to prepare proposals. You should not subject a consultant, contractor, or supplier to the hassle of preparing a proposal if they have little chance of getting the job. Some criteria for selecting consultants include:

- Experience
- Education
- References
- Subconsultants
- Project approach/new ideas
- Chemistry or fit with staff
- Financial health
- Insurance
- Familiarity with applicable facility systems and equipment
- Size of staff
- Years in business

Techniques for Team Decision-Making and Problem-Solving

Once the team is assembled and the project under way, there is an ongoing need for decisions, and for solutions to the problems that inevitably arise. Consensus decision-making is one of the most powerful tools at your disposal. Team decisions can minimize mistakes and disputes, and vastly increase productivity. It is a good idea to keep a few simple rules you can follow to diffuse conflicts between team members. For example:

- Affirm the opinions of the conflicting team members before expressing your perspective.
- Attack the problem or process, not the person's character.
- Speak the truth in a kind and respectful way.
- Separate facts from emotions.

The Function Analysis System Technique

The Function Analysis System Technique (FAST) applied in the value engineering process is a great example of team problem-solving. It gets a multi-disciplinary group of people together as a team to solve problems based on the analysis of functions. Function Analysis System Technique helps build a consensus among team members on what the problem is, how the problem will be solved, and why the selected corrective actions are being taken. This technique answers not only *how* something is being done, but *why*. Function Analysis System Technique is also a powerful tool to determine and rank priorities.

When this technique is applied in the value engineering process, customers can understand where their funds are being spent and can therefore decide which areas are most important when it comes to planning future uses for available funds. When an owner participates in the value engineering process, he or she will often seek trade-offs and set priorities that comply with budget parameters. This up-front communication helps avoid criticism when the project is under way or upon project completion, when "I wish I had done this or that" hindsight is common.

Cause and Effect Diagram

A practical tool for generating ideas and making decisions by consensus about a problem or issue is a Cause and Effect diagram, created by the project team. This is often called a "fishbone" diagram because its lines resemble the skeleton of a fish. The basic problem, issue, or desired effect becomes the "head" of the fish. Then the team identifies *causes* behind the basic problem or issue. Frequently the major causes are grouped into four categories: *manpower*, *methods*, *materials*, and *machinery*. However, teams may choose their own categories to identify and classify the major causes. During the process of creating the fishbone diagram, the team uses creativity techniques such as brainstorming to identify causes within each of the four categories, using a minimum number of words, which then become additional "bones" in each category. Sometimes teams include words that describe the environment; these become the "tail" of the fish. Use of the fishbone diagram can help organizations find new ways to improve and fine-tune their operations. An example is shown in Figure 10.1.

Affinity Diagram

An Affinity Diagram is another useful tool for team members to gather and classify shared ideas. An Affinity Diagram is effective in collecting information on a specific problem, as shown in Figure 10.2. This format can also be used with larger groups or when the topic offers a wide variety of choices that require grouping. Follow these steps to create an Affinity Diagram:

1. Define the issues the team is to consider.
2. Generate ideas individually, writing clearly on slips of paper or special sticky-back note paper.
3. Use a brief description for each idea.
4. Sort recorded ideas (silently) into related groupings by spreading them out on a table or posting them on a wall.
5. Create new categories using new or existing ideas.
6. Look for patterns and reach a team consensus on the highest ranking ideas.

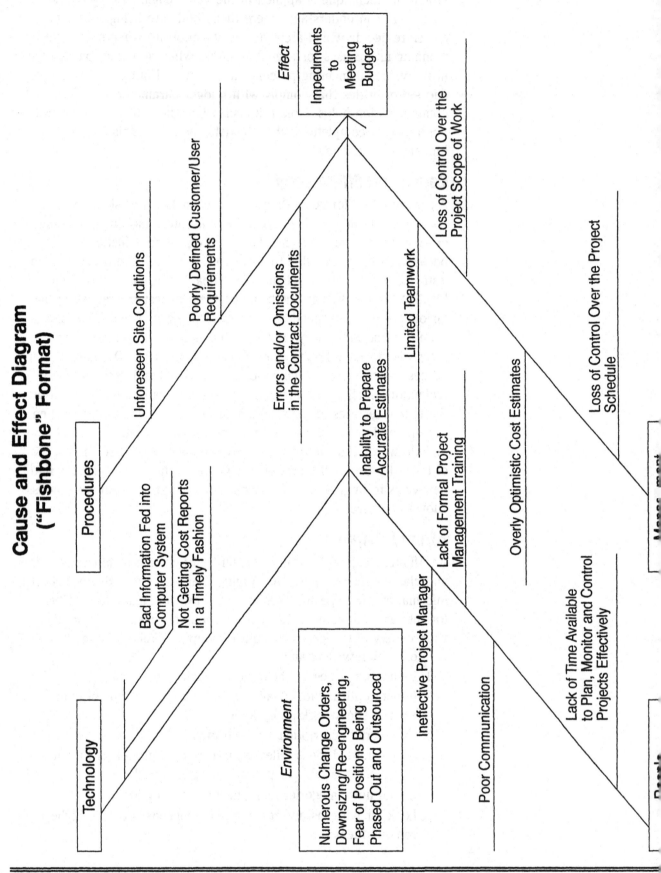

Cause and Effect Diagram ("Fishbone" Format)

Effect

Impediments to Meeting Budget

Procedures

Technology

Environment

Unforeseen Site Conditions

Poorly Defined Customer/User Requirements

Bad Information Fed into Computer System

Not Getting Cost Reports in a Timely Fashion

Errors and/or Omissions in the Contract Documents

Inability to Prepare Accurate Estimates

Limited Teamwork

Lack of Formal Project Management Training

Loss of Control Over the Project Scope of Work

Loss of Control Over the Project Schedule

Overly Optimistic Cost Estimates

Ineffective Project Manager

Numerous Change Orders, Downsizing/Re-engineering, Fear of Positions Being Phased Out and Outsourced

Poor Communication

Lack of Time Available to Plan, Monitor and Control Projects Effectively

Figure 10.1

Product Design	Material	Machine	Programming	Procedures
Panel width varies, requiring the rails to be set	Lack of sufficient part spec	Feeders broken	Lack of standard component library consistent over all lines	Lack of escalation policy
Large variation in board design	Purchasing looks at cost only	Method difficult to understand/complex	Inability to quickly convert programs	Lack of feeder procedure
Different part numbers for same hardware	No process to elevate design issues to engineering	Equipment out of calibration	Accountability issues on program changes	
Same part/different part	No incoming inspection	Power failures	Inconsistent CAD Files	
0 ohm resistors used	Material unavailable to run	Feeder maintenance		
No DFM	High rejection rates of parts	Non-standard equipment		
Extensive use of cutting edge				
Top/bottom side components				

Figure 10.2

Positive/Negative Forces Analysis

A two-column Positive/Negative Forces Analysis can also be used to stimulate thinking. For example, the format may be used to receive the team's input regarding the positive and negative aspects (or costs and benefits) of proposed alternative solutions. Other uses for this technique include analyzing the positive and negative aspects of a meeting at its conclusion, or to communicate what team members want, and do not want to occur on a project. Figure 10.3 is an example of positive/negative forces analysis used to determine the feasibility of various site locations for a new facility.

Post-Mortem Team Review

At the end of every significant project, it is helpful to assemble project team members to conduct a post-project evaluation. Ask yourselves "Knowing what we know now about the project, if we had to do it over again, what would we do differently?" Either the Positive/Negative

---

Site Analysis for the New Facility

	+	−
Alternate 1	**Pole Building Site** Ease of truck access Distant from utilities "Flat" site	Requires extensive demolition Visibility No expansion capability Eliminates expansion capability for adjacent building Distance to parking Requires new storage space
Alternate 2	**Parking Lot Site** Ease of truck access Adjacent parking Visibility Allows space for future expansion "Flat" site	Consumes existing parking space Distant from utilities
Alternate 3	**Addition to Existing Building Site** Part of campus Proximity to utilities Interconnects buildings More aesthetically pleasing	Distance to parking More difficult truck access More difficult construction access Expansion limited by setback requirements Relocation of high voltage feeders may be required Eliminates planned expansion for other operations

Figure 10.3   Positive/Negative Forces Analysis

Forces Analysis or Customer Attitude Survey form (Chapter 1) are useful tools for collecting this information.

In addition to significant planned projects, the post-mortem team review should also be applied to any project that requires emergency response from a contractor or in-house personnel. Taking a proactive stance and avoiding the necessity for work done under these circumstances represents a clear opportunity for savings because this kind of work often involves premium labor and equipment cost, and possible disruption of productivity.

Characteristics of Team Leaders

Decision-making and problem-solving skills are essential for productive facility management. Beyond this, managers need to make a lifelong commitment to cultivating teamwork and developing their own leadership abilities. In the pursuit of their profession, facility managers affect the quality of other people's lives. Conducting their work in an ethical manner is one of the fundamental ways they earn and maintain the confidence of team members, supervisors, peers, subordinates, suppliers, customers, and the general public.

The facility manager or project manager must not hide problems, acknowledging errors promptly, so that progress can be made toward a solution. Feeding bad information or withholding important information to try to control the situation is a good way to destroy credibility for the manager and his organization. Customers are entitled to accurate and timely facts so informed decisions can be made. Leaving out relevant information in status reports or misrepresenting potential costs to get a project approved can reduce the manager's credibility and effectiveness.

It is important to remember that the purpose of a team is to achieve an organization's objectives. A group of people might be inclined to choose someone with a dominant personality to become their team leader, with the idea that a take-charge type will get the job done. On the other hand, a domineering, authoritarian leader may lead the team down the wrong path. The best leaders want to serve and equip others.

Leaders who possess character value the people they lead and serve. Leaders of character think of themselves as servants, setting aside their own egos and self interests and inspiring other team members to act on behalf of the organization's greater goals. They demonstrate an appreciation for other team members' perspectives. They find ways to protect the interests of their followers and still achieve the collective goals of the organization.

Leaders should foster open, two-way communication. Commitment from team members is elicited by enabling them to participate jointly in analyzing problems and offering solutions. When a leader gives directions, he or she should provide opportunities for team members to ask questions in order to clarify what is expected of them. An effective leader also asks questions to uncover and resolve problems. He or she also reads body language to recognize signs of impatience, approval or disapproval, hesitance, confusion, or understanding.

Leaders must stay ahead of the game, including the competition. They strive to learn as much as they can from each personal and business situation. They understand the ultimate goal in learning is the application of that knowledge to overcome obstacles that will confront them in the future.

Effective leaders work to develop their own skill repertoire and empower others to expand theirs. One way to empower team members is by noticing their efforts and potential and encouraging them to persevere. Acknowledge each individual's accomplishments with a reward appropriate for his or her unique needs. Everyone needs to know that the work they do is appreciated and has merit.

To be an effective leader, one must have a vision of what needs to be accomplished. Leaders are goal-oriented. They plan their work and work their plan. Team members are often required to undertake projects that are unpleasant or difficult. They like to know what steps need to be taken to reach the objectives and how they will be rewarded for their efforts. Leaders motivate people to work toward a common goal.

Leaders must be good facilitators and use every meeting as an opportunity for team building. Sharing of experiences should be encouraged so team members can learn from and build on the knowledge of others. Leaders can tap into the knowledge and skills available through other people, empowering those individuals by initiating and facilitating team building workshops. Workshops may include group problem-solving activities such as Cause and Effect Analysis exercises, brainstorming, FAST diagramming, Affinity Diagramming, and Positive/Negative Forces Analyses.

Effective leaders are aware of their own strengths and weaknesses, as well as those of other team members. Even if every person on a team should possess similar skills, the differences in timing, behavioral patterns, communication preferences, and motivation affect each person's ability to work productively in various environments.

Complete the form in Figure 10.4 to become more aware of your unique passions, work preferences, skills, and habits. Share the completed form with key members of your team, for their impressions. Repeat this process with all team members periodically. By looking closely at yourself and others, you will take a leadership role in developing a more productive team. Each team member needs to understand how he or she helps or hinders the team as a whole. The process enables each team member to understand individual differences and to identify and set dates for making changes that will encourage the effectiveness of the individual and the work team.

What do you like? (List work activities or areas that you enjoy doing or have a passion for.)

_____

_____

_____

What do you dislike? (Identify those work activities that are less appealing or that you tend to avoid.)

_____

_____

_____

Solicit input from your team members on the following items. Be fair, honest, and diplomatic in this exchange.

What are you good at? (List special knowledge, talents, and skills.)

_____

_____

_____

What skills do you need to learn, improve, or develop?

Skills   Target Completion Date:

_____   _____

_____   _____

_____   _____

What are your best work habits?

_____

_____

What present work habits do you need to change, or what new habits do you need to develop?

_____

_____

Habits   Target Completion Date:

_____   _____

_____   _____

_____   _____

Figure 10.4   Team Member Awareness Form

What are your working style tendencies? (Use the rating scale below for assistance in clarifying your style. Circle the number on the scale that best describes you.)

Structured   Unstructure
1   2   3   4   5   6   7   8   9   1

Decisive   Indecisiv
1   2   3   4   5   6   7   8   9   1

Proactive   Reactiv
1   2   3   4   5   6   7   8   9   1

Patient   Impatie
1   2   3   4   5   6   7   8   9   1

Organized   Disorganize
1   2   3   4   5   6   7   8   9   1

Concise   Expansiv
1   2   3   4   5   6   7   8   9   1

Reserved   Outspoke
1   2   3   4   5   6   7   8   9   1

Detail-Oriented   Concept-Oriente
1   2   3   4   5   6   7   8   9   1

Impetuous   Delibera
1   2   3   4   5   6   7   8   9   1

People-Oriented   Technically-Oriente
1   2   3   4   5   6   7   8   9   1

Time-Oriented   Results-Oriente
1   2   3   4   5   6   7   8   9   1

Aggressive   Passiv
1   2   3   4   5   6   7   8   9   1

Figure 10.4   Team Member Awareness Form cont.

Epilogue: Where Do You Go From Here?

The challenge of corporate facility management is to facilitate the most economic and efficient use of facilities in response to competitive threats and opportunities. Resources for replacement, development, and maintenance of facilities are scarce in most organizations. The appropriate concepts, strategies, techniques and skills must be applied in a timely and systematic manner to optimize the return on facility assets and meet constantly changing customer demands. When the concepts and methods of Total Productive Facility Management are understood and implemented in a sound way, they can help managers do more for less and to do it more quickly.

Power is not necessarily bestowed upon those with the most money, brains, or academic degrees. Rather it abides in the individual or organization people turn to for competent advice and help. Facility managers will achieve power and recognition by helping others do their jobs better and faster—and by facilitating the teamwork necessary for productive and safe working environments. Now you are ready to go!

The first step, as stated in Part 1, "Evaluating Facilities Performance and Setting Goals," is to acknowledge the need for change and commit to a process of ongoing learning and improvement. To provide the basis for development of strategies that will benefit both the facility manager and his or her organization, it is necessary to clarify the greater goals of the company on a continuing basis. Assess facilities performance from the perspectives of internal and external customers and compare your performance against your competition and industry best practices.

Business situation analyses, customer attitude surveys, space utilization studies, flow charts, benchmarking, and value engineering are practical methods for collecting, organizing, and analyzing data; identifying performance requirements; and determining the developmental changes needed to keep the entire organization healthy

231

and vigorous. These methods incorporate multidisciplinary teams, cross-functional systems thinking, and creativity techniques to foster cooperation and consensus decision-making. Remember to evaluate facility management's contribution to corporate goals on a routine basis, because the performance targets and priorities are always changing.

Meticulously apply value engineering to define the functions of facilities, projects, and services This will help determine lower cost ways to reliably provide the required functions. Customers pay for functions. Value engineering determines which functions and services are the most valuable and how much they are worth from the customers' perspective. Facility management organizations should be prepared to justify the value of their projects and services at all times. If the facility manager does not initiate the value studies, someone else in upper management may. Be proactive!

As stated in Part 2, "Project-by-Project Improvement," the next step is to create a project-driven facility management organization and employ a structured project management approach for implementing the desired changes. Business and financial conditions, regulatory requirements, technology developments, and competition influence the type of projects selected. In a competitive, resource-constrained environment, there will always be emphasis on projects that help improve the organization's cost, quality, and cycle-time management.

Matrix organizational structures, decision charts, work breakdown structures, programming and scope documents, cost estimates, CPM schedules, value engineering, responsibility charts, life cycle cost analysis and selection of the appropriate contracting and procurement methods are essential for effective project management. Supervision use of feedback systems and reports, and visible, easily observable milestones are necessary to measure and control project progress. Some of the benefits of structuring an organization around the management of projects include: efficient allocation and control of limited resources, shorter development times, minimum cost, and improved quality and performance.

The final step is to improve productivity through communications and teamwork, as described in Part 3. Many organizations can dramatically improve effectiveness and productivity by making improvements in these areas. Management alone cannot keep a company competitive anymore. Everyone needs to share a sense of responsibility for the survival and success of the company.

Today, most organizations are unwilling to empower a single individual to make high risk decisions that impact the competitiveness and return on expensive facility assets. The high stakes, increased number of stakeholders, and the complexity of required tasks, skills, technology and knowledge require the involvement of a diverse mix of people from inside and outside the organization.

The broad task of the facility manager is to facilitate the creation of flexible teams that can manage projects and respond effectively to new workplace challenges. Facility managers need to work continuously to improve team effectiveness, share ideas, and develop esprit de corps. Many individuals who have been informed that they are critical components of a team do not feel that they are really working together as a team to solve shared problems in the most efficient and effective manner. Changing current methods of working and organizing company resources takes teamwork, team building, team incentives, and shared information. Conduct team-building workshops using tools such as cause and effect diagrams, affinity diagrams, and positive and negative forces analyses.

Facility managers should focus attention away from building their own power bases and refocus on completion of tasks that will improve the entire company. A fluid and efficient business organization requires timely and effective communication. Through telephones, pagers, e-mail, facsimiles, and express deliveries, distance is practically eliminated as a barrier to communication. Dissemination of the right information to the right individuals at the right time improves the probability that a project will be successful.

Create an ongoing communication system to inform the entire business enterprise about the services provided by the facility management department. Some practical communication tools include an internal brochure, newsletter, bulletin boards in common areas, and e-mail bulletin board messages. Make a commitment to facilitate good communications.

The best facility managers are equipped with a variety of methods which they continually fine-tune and adapt to fit the specific needs of each situation. They have control over both the cost and productivity of facility resources and foster efficient and effective working environments in which their team will flourish and function well.

Appendix

Bibliography/
References

Index

Appendix

Table of Contents

Appendix A
Value Analysis Study: Sample Project · 238

Appendix B
Conceptual Construction Cost Estimate: Sample Project · 243

Appendix C
Factors for Annual Compounding Interest · 247

237

Final Report of the Value Analysis Study for a Telephone Utility Company Central Office Remodeling Project

Project No 3693
Project Hours: 80

VALUE ANALYSIS TEAM MEMBERS

Team Leader. Value Engineering Specialist
                Client Representative, Telephone Utility Company
                Client Representative, Telephone Utility Company
                Client Representative, Telephone Utility Company
                Construction Manager
                Construction Manager
                Mechanical Engineer
                Electrical Engineer
                Architect

Project Description The project consists of remodeling approximately 13,000 square feet of space in an existing two-story building of approximately 37,000 square feet with a partial basement. Some telephone switching equipment space and existing office-type space will be converted to accommodate operator workstations, conference rooms, training rooms and related support areas.

Scope of Value Analysis Study The value analysis focused on evaluating certain HVAC system retrofit options relative to their ability to minimize costs to the building owner over the life of the building system. HVAC elements under the scope of the study included primarily air distribution and heating options (The refrigeration system was not reviewed because any savings that would be generated would be outweighed by the incremental retrofit cost a new system would represent when compared with an existing system. Value engineering should have been applied to the refrigeration system during the initial design of the system in 1977.)

Computer models of the building were created utilizing energy analysis software. These models took into account the building's physical construction, indoor design conditions, lighting systems, projected occupancy levels and schedules, electrical equipment and HVAC equipment's operating characteristics. (Much of this information was provided by the telephone utility company, the architect, and the mechanical and electrical engineers) The computer models generated annual electrical and gas consumption and demand figures on a system-by-system basis. The energy consumption figures for the models were compared, energy savings were established, and the associated cost savings were calculated. These savings and projected construction costs were then used to calculate simple payback periods for the system options that were evaluated.

Model Descriptions The first model was based on existing building construction and architects' and engineers' preliminary retrofit design recommendations. Features of this model included the following:
- 1 5 Watts/Sq. Ft. assumed for air conditioning loads from building lighting
- Electric reheat coils to provide building heating and reheating of minimum air flow quantities
- VAV box minimum airflow settings based on 40% of nominal box capacity
- Electric heating of vestibules and second floor utility spaces

The second model was based on a composite of the engineering firm's design options for reducing system operating costs. (Note that the models generated with energy analysis software simulate system energy usage and demand only The associated energy costs and estimated maintenance costs were manually calculated at the conclusion of the computer modeling.) Features of the value engineering model included the following:
- 1 00 Watt/Sq. Ft. used as maximum lighting load in all remodeled areas. (Based on use of electronically ballasted fluorescent light fixtures with deep cell parabolic lenses, arranged to provide the minimum project illumination criteria of 30 to 40 footcandles.)
- Re-use of existing steam baseboard heating system for perimeter heating requirements, including the east and west entry vestibules and the "intermediate floor."
- Elimination of VAV box minimum airflow settings and associated electric heating/reheating coils.(Based on allowing VAV boxes serving interior spaces to fully close when no cooling is required, assuming that when space is occupied, cooling will be required and VAV boxes will be open enough to admit code-required air.)
- Replacement of existing standard motors at supply and return/exhaust fans with high efficiency motors.

deling Results The results of the modeling calculations are attached for detailed review, but the summary
ow collates and compares the results of the models, and presents dollar figures associated with the calculated
ergy savings.

Table 1 indicates, operating cost savings can be attained in several of the building's energy-consuming
mponents. The lighting cost savings of $2,537 are a direct result of using more energy-efficient lighting methods
reduce energy consumption to a maximum of 1.0 Watt/Sq Ft. Through the use of electronic ballasts and deep-
l parabolic fixtures, the electrical designer should be able to attain the design illumination level of 30 to 40
tcandles with 1 0 Watt/Sq. Ft. or less for lighting. However, meeting this criteria will depend upon aspects of
e design other than the fixture types alone. It will also depend on ceiling heights and colors used in the furniture
d architectural finishes. But if a cooperative approach is adopted by the design team, the conditions outlined
ove should be achievable.

le 1 indicates that through the use of the existing baseboard heating system in lieu of the proposed VAV
stem with electric reheat, annual energy savings of $3,649 will be generated This is primarily due to the
mination of minimum CFM settings at the VAV boxes, not the use of one heating source in lieu of another. The
imum CFM requirements result in airflow to a space when no cooling is required. If no cooling is required and

Table 1
Telephone Utility Company–Value Engineering Analysis, Central Office

Building Component	Utility Description	Engineers' Proposed Design	Value Engineering Composite Design	Energy Demand Reductions	Annual Savings
Lighting	Annual kWh	215,707 00	174,904 00	40,803 00	$1,632 00
	Summer Demand	172 00	149 60	23 20	353 00
	Non-Summer Demand	345 60	299 20	46 40	552 00
				Lighting Savings Subtotal	**$2,537.00**
Electric Heating	Annual kWh	37,710 00	—	37,710 00	$1,508 00
	Summer Demand	48 40	—	48 40	736 00
	Non-Summer Demand	120 40	—	120 40	$1,432 00
Steam Baseboard Heating	Annual Therms	—	72 00	(72 0)	($26 00)
	Annual kWh	—	8 00	(8 00)	(0 00)
	Summer Demand	—	0 00	0 00	0 00
	Non-Summer Demand	—	0 10	(0 10)	(1 00)
				Heating Equip. Subtotal	**$3,649.00**
Supply Fan	Annual kWh	438,956 00	376,889 00	62,067 00	$2,483 00
	Summer Demand	314 90	294 00	20 90	318 00
	Non-Summer Demand	536 80	495 20	41 60	495 00
				Supply Fan Savings Subtotal	**$3,295.00**
Return Fan	Annual kWh	92,641 00	76,389 00	16,252 00	$650 00
	Summer Demand	68 80	61 30	7 50	114 00
	Non-Summer Demand	115 40	102 10	13 30	158 00
				Return Fan Savings Subtotal	**$922.00**
Chiller & Refr. Equipment	Annual kWh	229,621 00	218,311 00	11,310 00	$452 00
	Summer Demand	491 60	491 70	(0 10)	(2 00)
	Non-Summer Demand	241 60	237 30	4 30	51 00
				Refr. Equipment Savings Subtotal	**$502.00**
				Total Annual Savings	**$10,905.00**

air is being forced into a room, that air must be heated to room temperature to prevent overcooling. Under the initial design concept, this heat is provided by electric coils in the VAV terminals. The Value Engineering design calls for the elimination of minimum airflow settings at the VAV terminals, thus eliminating the need for reheat coils since space heating is provided by steam-heated baseboard units and duct coils. And based on the lighting, occupancy and internal equipment utilization schedules used in the computer models, very little space heating is required in addition to the heat generated internally.

The indicated supply and return fan savings are significant. But the energy and cost savings indicated in Table 1 do not result from the motor change-out alone. The reduction in lighting levels in some areas reduces the lighting load, and thus the amount of air the fans must move is reduced. This in turn reduces the energy consumption at the supply and return fans. Similarly, the elimination of minimum airflow settings at the VAV terminals reduces annual fan energy consumption by allowing the VAV terminals to go fully closed when space cooling is not required. Thus, the indicated fan energy savings is a composite of the fan savings due to reductions in lighting loads, elimination of VAV terminal minimum settings, and use of high efficiency fan motors.

Finally, the chiller and refrigeration equipment operating cost reductions are the results of all of the modifications listed above. The reduction in lighting levels reduces the cooling load contribution made by the lights. The elimination of minimum airflows at the VAV terminals reduces the airflow at the cooling coil when cooling is not required in specific areas, thus reducing the overall cooling load. The high efficiency fan motors reject less motor heat to the air stream than standard motors. The resulting cooling load reductions decrease the amount of work the refrigeration equipment must perform, generating an overall refrigeration system energy cost savings of $502.

As Table 1 indicates, if all of these options are implemented, the total annual energy savings is $10,905.

Value Analysis The information in Table 2 extends the value analysis to include the impact of first costs and maintenance costs. For the lighting evaluation, it was assumed that first costs and maintenance costs for the two options would be approximately equal and therefore have no impact.

The heating system evaluation takes into account demolition of the existing radiators, inclusion of electric reheat coils at the VAV terminals, and providing power to the reheat coils under the engineer's originally proposed design. For the value engineering design, the evaluation includes the cost of architectural enclosures for the radiators, and the installation and maintenance costs associated with providing a smaller, more energy efficient boiler for the steam heating system. Both evaluations assume that the existing boiler is abandoned in place. The value engineering design also assumes that the existing boiler ancillaries are retained and reused for the new boiler. The value engineering design costs more to install ($22,800), but results in $3,149 annual savings. The resultant simple payback period is just over seven years.

One alternative to the value engineering design would be to retain the existing boiler in lieu of providing a new unit. But annual maintenance costs for the larger, older boiler have been estimated at 6 times that of the new, small boiler. The results are that the first costs are reduced below that of the VAV with electric reheat system ($3,200 savings), but annual operating cost savings are reduced to $649. Careful reevaluation of the original design, with an emphasis on eliminating reheat coils and reducing VAV terminal minimum airflows where possible could reduce the first costs and operating costs indicated for the VAV-reheat design to a point where the difference between the two designs is insignificant. At that point, the decision whether to use VAV with reheat, or steam baseboard heat becomes an evaluation of intangibles that the owner must undertake. VAV with reheat offers better air circulation when little or no cooling loads are present in the space. Steam baseboard heat offers future flexibility to convert from one energy source to another (gas to oil to electric) depending on the prevailing energy market.

(Note. During a meeting between the Owner, Architect, Engineer, Contractor and Value Engineer, the various aspects of the two heating system options were discussed. Due to the age of the existing steam supply and condensate return system, the need for some radiators to be relocated, and a likely conflict between the existing steam and condensate piping in the crawl space with proposed duct modifications, the analysis team agreed that VAV with electric reheat should be implemented, with VAV box minimum airflows and reheat coil kW capacities to be reevaluated and reduced where possible.)

The figures in Table 2 related to fan motor replacement appear to indicate a simple payback of only 8.5 months ($3,000 cost to achieve $4,217 annual savings). But, as described previously, the fan operation savings are due to more than just replacing the motors. The annual energy savings solely attributable to the fan motor change-out

Table 2
Value Engineering Analysis

Engineer's Proposed Design

System/Component	Installation Costs	Operation	Maintenance	Total
Lighting at 1 5 Watts/S F in Remodeled Office Areas (Operating Costs for Lighting Only)	—	$15,366	—	$15,366
Radiator Demolition	$15,800	—	—	—
Additional Costs Assoc with Providing Reheat Coils for VAV Boxes				
— Cost of Reheat Coils	$5,000	—	—	—
— Electrical Power	$11,400	$3,676	—	$3,676
HEATING SUBTOTAL	$32,200	$3,676	$0.00	$3,676
Radiator Demolition	$15,800	—	—	—
Additional Costs Assoc with Providing Reheat Coils for VAV Boxes				
— Cost of Reheat Coils	$5,000	—	—	—
— Electrical Power	$11,400	$3,676	—	$3,676
HEATING SUBTOTAL	$32,200	$3,676	$0 00	$3,676
Additional Miscellaneous Items				
Existing Fan Motors	—	—	—	—
— Supply Fans	—	$28,730	—	$28,730
— Return Fans	—	$6,124	—	$6,124
FAN MOTORS SUBTOTAL	—	$34,854	—	$34,854
Existing Refrg Equip	—	$19,535	—	$19,535

Proposed Value Analysis Alternative

System/Component	Installed Costs	Operation	Maintenance	Total	Savings First Cost	Savings Annual
Lighting at 1 0 Watt/S F in Remodeled Office Areas (Operating Costs for Lighting Only)	—	—	—	$12,823	$0.00	$2,537
Radiator Enclosures	$29,000	—	—	—	—	—
New Gas-Fired Steam Generator for Radiators to Reman (VAV Terminals Set to O CFM Minimum Airflow)						
— Electrical Power	$26,000	$27	$500	$527		
HEATING SUBTOTAL	$55,000	$27	$500	$527	($22,800)	$3,149
Radiator Enclosures	$29,000	—	—	—	—	—
Existing Steam Boiler Retained and Reused for Space Heating						
— Electrical Power		$27	$3,000	$3,027		
HEATING SUBTOTAL	$29,000	$27	$3,000	$3,027	$3,200	$649
Additional Miscellaneous Items						
Operating Cost Reductions for the Following Items Due to a Combination of Factors, Including but not Limited to Motor Replacement						
New, High-Effic Fan Motors	$3,000	—	—	—		
— Supply Fans	—	$25,435	—	$25,435		
— Return Fans	—	$5,202	—	$5,202		
FAN MOTORS SUBTOTAL	$3,000	$30,637	—	$30,637	($3,000)	$4,217
Existing Refrg Equip	—	$19,033	—	$19,033	$0 00	$502

are approximately $1,780. While this is far less than $4,217, the revised simple payback period (20.2 months) is still attractive.

The refrigeration system operating cost reductions did not require any retrofit, nor was there a change in the annual maintenance costs associated with it. Therefore, no independent evaluation or discussion is required for this system. The tabulated energy cost savings simply enhance the overall benefit derived from the other modifications.

HVAC Value Study Estimate of Savings

Recommendation	Annual Energy Savings
1. Reduce lighting energy from 1.5 watts per square foot to 1 watt per square foot in certain areas of the facility to reduce lighting and HVAC operating costs.	$9,857
2. Change variable air volume minimum airflow settings to zero when spaces are unoccupied.	$1,432
3. Change standard supply and return fan motors on air handling units to energy-efficient motors.	$2,650
4. Install an outside air economizer on unit S-2.	$10,123
Total	**$24,062**

Additional V.E. Recommendations Recorded in Meeting Minutes

Conference Report

1. The existing boiler will no longer be utilized and will be retired-in-place. Retention or replacement with a new boiler is not cost-effective

2. Electric reheats will be used for the new project. Energy will be minimal due to light demand resulting from internal heat gains.

3. Existing radiator piping will be removed in the crawl space to eliminate conflicts with new vertical duct shafts

4. No heating will be required for the first and second floor equipment areas.

5. Electric heat will be provided to replace the radiator in the intermediate mezzanine.

6. Reheat units will be set at a minimum opening (not necessarily zero) to conserve energy.

7. The mechanical and electrical engineer has been requested to rework ductwork for the existing satellite air conditioning unit in the equipment area and modify the existing VAV units.

8. The mechanical and electrical engineer will contact the electrical company to investigate rate adjustment incentives for going to all electric heating

9. After analysis, it was determined to design building mechanical systems around equipment requirements and not to utilize the existing satellite air conditioning unit for supplemental air.

10. The mechanical and electrical engineer will notify the construction manager where existing piping will remain.

11. The two air handling supply and two return fan motors will be replaced with high-efficiency motors

SAMPLE PROJECT
CONCEPTUAL CONSTRUCTION COST ESTIMATE

ESTIMATE BASED ON *MEANS BUILDING CONSTRUCTION COST DATA**

	Qty	Unit of Measure	Price Per Unit of Measure	Total Price	Contractor's Estimates Price Per Unit of Measure	Total Price
DIVISION 1 - GENERAL REQUIREMENTS						
100 General Requirements				4,677 74		4,677 74
166 Trash Removal	5	each	335 00	1,675 00	317 25	1,586 25
110 Final Cleaning	14,500	sq ft	0 15	2,175 00	0 14	2,030 00
DIVISION 2 - SITE WORK						
272 Demolition for Remodeling						
1 Remove existing 9' lay-in ceiling	4,736	sq. ft	0 38	1,799 68	0 19	899 84
2. Remove existing countertop	70	lin ft	3 82	267 40	0 71	49 70
3 Remove existing base cabinet	6	lin ft	5 71	34 26	12 63	75 78
4 Remove existing 9' high partition	1,791	sq ft	0.26	465 66	0 83	1,489 53
5. Remove existing base cabinet	12	lin. ft	5 71	68 52	8 35	100 20
6 Remove existing countertop	24	lin. ft	3 82	91 68	2 09	50 16
7 Remove existing 16' high drywall partition	1,888	sq ft.	0 26	490 88	0 63	1,189 44
8 Remove existing toilet partition	120	sq ft	0 76	91 20	1 05	126 00
9 Remove existing carpet	4,484	sq ft	0.23	1,031 32	0 15	672 60
10 Remove existing 6' x 9' double door and metal frame	3	each	50 46	151 38	25 03	75 09
11 Cut 3' x 9' opening for relocation of door into existing drywall partition	1	each	65 69	65.69	93.88	93 88
12 Cut 6' x 9' opening for new door into existing drywall partition	1	each	131.38	131 38	93 88	93 88
13 Remove existing 5' long vanities	10	sq ft	5 71	57 10	10 01	100 10
14. Remove existing toilet partitions	120	sq ft	0 76	91 20	0 63	75 60
15 Remove existing concrete slab	210	sq ft	12 61	2,648 10	2.02	424 20
200 Paving and Surfacing						
1 Fill in exterior loading dock ramp (Allowance)				5,000 00		5,000 00
DIVISION 3 -CONCRETE						
300 Cast-In-Place Concrete						
1 Replace concrete floor slab	4	cu yd	49 28	197 12	105 90	423 60
DIVISION 4 - MASONRY Not Used						
DIVISION 5 - METALS Not Used						
DIVISION 6 - WOOD AND PLASTICS						
100 Rough Carpentry						
1 Remove existing 4' x 3' metal frame and glass, fill in and patch	4	each	58 99	235 99	15 63	62 52
2 Fill and patch 6' x 9' drywall opening	54	sq ft	2 14	115 56	2 32	125.28
3 9' high drywall partition with 5/8" drywall on both sides	2,133	sq ft	2.14	4,564 62	2 48	5,289 84

		Qty	Unit of Measure	Price Per Unit of Measure	Total Price	Price Per Unit of Measure	Total Price
4	16' high drywall partition with 5/8" drywall on both sides	2,432	sq ft	2 14	5,204 48	2 48	6,031 36
5	10' high drywall partition with 5/8" drywall on both sides	1290	sq ft	2 14	2,760 60	2 48	3,199 20
6	Furr out four columns with metal stud and 5/8" drywall	144	sq ft	2 14	308 16	2 60	374 40
7.	Patch drywall and ceramic tile at wall at new lavatory in existing women's toilet room	16	sq ft	4 63	74 08	1 89	30.24
8	Remove 5/8" drywall to insulate with 4" Sound Batt Insulation	288	sq. ft	0 51	146 88	0 25	72 00
9	Install 5/8" drywall which had been removed to insulate with 4" Sound Batt Insulation	288	sq. ft	0 73	210 24	0 47	135 36

6220 Millwork

1	Four 5'-6" plastic laminated countertops	22	lin ft	24.04	528 88	37 55	826.10
2	Plastic laminated countertops	8	lin. ft	24 04	192 32	40 68	325.44

6410 Cabinetwork

1	Base Cabinet	1	each	169 52	169 52	325 43	325 43
2	Base Cabinets	2	each	147 52	295 04	325 43	650 86
3	Wall Cabinets	4	each	118.99	475.96	162.79	651 16

DIVISION 7 - THERMAL AND MOISTURE PROTECTION

7200 Insulation

1	4" Sound Batt Insulation for plant	2,432	sq ft.	0 36	875 52	0 42	1,021.44
2	4" Sound Batt Insulation for office	1,188	sq ft	0 36	427 68	0 42	498 96
3.	4" Sound Batt Insulation for ceiling	768	sq ft	0 83	637.44	0 44	337 92

7500 Roofing

1	Roof repair at 7 locations (Allowance)	1,148	sq. ft	1 04	1,193 92	1 04	1,193 92

DIVISION 8 - DOORS AND WINDOWS

8200 Doors and Frames

1	Remove and relocate 3' x 9' wood doors and metal frames	11	each	88 82	977 02	118 91	1,308 01
2	6' x 8' double metal door and frame with 2' x 2'-6" light in each door	4	each	15 35	61 40	938 74	3,754 96
3	3' x 7' metal door and frame	2	each	840 32	1,680 64	481 88	963 76
4	3' x 9' metal frame with cased opening	1	each	344 31	344.31	93.88	93.88
5	5' x 7' bifold doors	2	each	607 78	1,215.56	575 57	1,151 14

8400 Entrances and Storefronts

1.	Emergency exit door and sidewalk (Allowance)				6,000 00		6,000.00

8500 Metal Windows

1.	2' x 9' side lights and metal frames	4	each	532 93	2,131.72	312.92	1,251 68

8700 Hardware

1	Replace hardware on existing relocated doors (Allowance)	11	each	200 00	2,200.00	200.00	2,200.00
2	Set of hinges for double action doors (Allowance)	1	each	1,000 00	1,000 00	1,000 00	1,000 00

	Qty	Unit of Measure	Price Per Unit of Measure	Total Price	Price Per Unit of Measure	Total Price
30 Mirror Glass						
1 Two 5'-6" x 4' mirrors	44	sq ft	18 71	823 24	8 70	382 80
VISION 9 - FINISHES						
10 Ceramic Tile						
1 Ceramic tile floor for toilet rooms	260	sq ft.	6 56	1,705 60	7 51	1,952 60
2 Ceramic tile 9' high wall in new toilet room	342	sq ft	4 87	1,665 54	7 51	2,568 42
60 Plastic Tile						
1 Vinyl composite tile for production, kitchen, and storage areas	8,988	sq ft	1 27	11,414 76	1 37	12,313 56
2 Vinyl base for production area	776	lin ft	1 52	1,179 52	1 25	970 00
3 Vinyl base for conference rooms	83	lin ft.	1.52	126 16	1 25	103 75
4 Vinyl base for toilet rooms	124	lin ft	1 52	188 48	1 25	155 00
5 Vinyl base for office area	784	lin ft	1.52	1,191 68	1 25	980 00
10 Acoustical Ceiling						
1 New 2' x 2' lay-in ceiling, 9' high	6,800	sq ft	2 96	20,128 00	1 40	9,520 00
2 New 2' x 4' mylar-faced cleanable lay-in ceiling, 10' high	8,074	sq ft	1 98	15,986 52	1 44	11,626 56
80 Carpet						
1 Carpet for office area	590	sq. yd	25 93	15,298 70	13 77	8,124 30
20 Interior Painting						
1 Paint metal door frames	27	each	34 36	927 72	50 07	1,351 89
2 Paint production area	7,760	sq ft	0 49	3,802 40	0 34	2,638 40
3 Paint office	7,056	sq ft	0 49	3,457 44	0 31	2,187 36
50 Wall Covering						
1 Wall covering for conference rooms	747	sq ft	0 94	702 18	1 51	1,127 97
2 Wall covering for toilet rooms	1,116	sq ft	0 94	1,049 04	1 51	1,685 16
VISION 10 - SPECIALTIES						
600 Partitions						
1 Urinal screen	2	each	272 18	544 36	150 19	300 38
2 Handicapped toilet stall partitions	2	each	595 92	1,191 84	719 70	1,439 40
3. Standard toilet stall partitions	4	each	415 43	1,661 72	657 11	2,628.44
800 Toilet and Bath Accessories						
1 Toilet paper dispenser	6	each	33 90	203 40	31 29	187 74
2 Sanitary napkin dispenser	2	each	396.33	792 66	375 49	750.98
3 Sanitary napkin disposal	4	each	396 33	1,585 32	50 07	200 28
4 3" x 1-1/2" grab bars	2	each	51 09	102 18	56 33	112 66
5 4" x 1-1/2" grab bars	2	each	68 12	136 24	56 33	112 66
6 Paper towel dispenser	2	each	48 23	96 46	56 33	112 66
7 Paper towel disposal	2	each	179 54	359 08	56 33	112 66
900 Wardrobe Specialties						
1 Wall shelf and rod	12	lin ft	5 01	60 12	18 78	225 36

VISION 11 - EQUIPMENT
Not Used

VISION 12 - FURNISHINGS
Not Used

VISION 13 - SPECIAL CONSTRUCTION
Not Used

	Qty	Unit of Measure	Price Per Unit of Measure	Total Price	Price Per Unit of Measure	Total Price
DIVISION 14 - CONVEYING SYSTEMS						
Not Used						
DIVISION 15 - MECHANICAL						
15400 Plumbing						
1 Plumbing				20,010 00		20,010 00
15500 Fire Protection						
1 Fire protection				63,250.00		63,250 00
15800 Air Distribution						
1 Ventilation work				59,015 00		59,015 00
2 Rooftop units				25,300 00		25,300 00
15900 Controls and Instrumentation						
1. Temperature controls (Allowance)				3,900 00		3,900.00
DIVISION 16 - ELECTRICAL						
16000 Electrical						
1. New 800 Amp 480 Volt service to building distribution gutter				3,000 00		3,000.00
2. New service to tenant space, power distribution equipment and light fixtures				36,000 00		36,000.00
3 Tenant space electrical distribution (wire, conduit, etc) (Excluding distribution to production equipment)				<u>12,000 00</u>		<u>12,000 00</u>
SUBTOTAL				364,093 18		344,451 44
Contingency (10%)				36,409 32		34,445.14
SUBTOTAL				400,502 50		378,896 58
Architectural/Engineering				28,000 00		28,000 00
Construction Management				25,000 00		25,000.00
Site Supervision and Reimbursable Expenses				<u>19,534 00</u>		<u>19,534 00</u>
TOTAL CONCEPTUAL CONSTRUCTION COST ESTIMATE AS INDICATED				**$473,036.50**		**$451,430.58**

* Adjusted by division, using the 1994 Houston City Cost Index (except Divs 1, 15, 16 allowance areas)

n	Single Payment Compound-Amount Factor (F/P,i,n)	Present-Worth Factor (P/F,i,n)	Equal-Payment Series Compound-Amount Factor (F/A,i,n)	Sinking-Fund Factor (A/F,i,n)	Present-Worth Factor (P/A,i,n)	Capital-Recovery Factor (A/P,i,n)	Uniform Gradient-Series Factor (A/G,i,n)
1	1.050	0.9524	1.000	1.0000	0.9524	1.0500	0.0000
2	1.103	0.9070	2.050	0.4878	1.8594	0.5378	0.4878
3	1.158	0.8638	3.153	0.3172	2.7233	0.3672	0.9675
4	1.216	0.8227	4.310	0.2320	3.5460	0.2820	1.4391
5	1.276	0.7835	5.526	0.1810	4.3295	0.2310	1.9025
6	1.340	0.7462	6.802	0.1470	5.0757	0.1970	2.3579
7	1.407	0.7107	8.142	0.1228	5.7864	0.1728	2.8052
8	1.477	0.6768	9.549	0.1047	6.4632	0.1547	3.2445
9	1.551	0.6446	11.027	0.0907	7.1078	0.1407	3.6758
10	1.629	0.6139	12.587	0.0795	7.7217	0.1295	4.0991
11	1.710	0.5847	14.207	0.0704	8.3064	0.1204	4.5145
12	1.796	0.5568	15.917	0.0628	8.8633	0.1128	4.9219
13	1.866	0.5303	17.713	0.0565	9.3936	0.1065	5.3215
14	1.980	0.5051	19.599	0.0510	9.8987	0.1010	5.7133
15	2.079	0.4810	21.579	0.0464	10.3797	0.0964	6.0973
16	2.183	0.4581	23.658	0.0423	10.8378	0.0923	6.4736
17	2.292	0.4363	25.840	0.0387	11.2741	0.0887	6.8423
18	2.407	0.4155	28.132	0.0356	11.6896	0.0856	7.2034
19	2.527	0.3957	30.539	0.0328	12.0853	0.0828	7.5569
20	2.653	0.3769	33.066	0.0303	12.4622	0.0803	7.9030
21	2.786	0.3590	35.719	0.0280	12.8212	0.0780	8.2416
22	2.925	0.3419	38.505	0.0260	13.1630	0.0760	8.5730
23	3.072	0.3256	41.430	0.0241	13.4886	0.0741	8.8971
24	3.225	0.3101	44.502	0.0225	13.7987	0.0725	9.2140
25	3.386	0.2953	47.727	0.0210	14.0940	0.0710	9.5238
26	3.556	0.2813	51.113	0.0196	14.3752	0.0696	9.8266
27	3.733	0.2679	54.669	0.0183	14.6430	0.0683	10.1224
28	3.920	0.2551	58.403	0.0171	14.8981	0.0671	10.4114
29	4.116	0.2430	62.323	0.0161	15.1411	0.0661	10.6936
30	4.322	0.2314	66.439	0.0151	15.3725	0.0651	10.9691
31	4.538	0.2204	70.761	0.0141	15.5928	0.0641	11.2381
32	4.765	0.2099	75.299	0.0133	15.8027	0.0633	11.5005
33	5.003	0.1999	80.064	0.0125	16.0026	0.0625	11.7566
34	5.253	0.1904	85.067	0.0118	16.1929	0.0618	12.0063
35	5.516	0.1813	90.320	0.0111	16.3742	0.0611	12.2498
40	7.040	0.1421	120.800	0.0083	17.1591	0.0583	13.3775
45	8.985	0.1113	159.700	0.0063	17.7741	0.0563	14.3644
50	11.467	0.0872	209.348	0.0048	18.2559	0.0548	15.2233
55	14.636	0.0683	272.713	0.0037	18.6335	0.0537	15.9665
60	18.679	0.0535	353.584	0.0028	18.9293	0.0528	16.6062
65	23.840	0.0420	456.798	0.0022	19.1611	0.0522	17.1541
70	30.426	0.0329	588.529	0.0017	19.3427	0.0517	17.6212
75	38.833	0.0258	756.654	0.0013	19.4850	0.0513	18.0176
80	49.561	0.0202	971.229	0.0010	19.5965	0.0510	18.3526
85	63.254	0.0158	1245.087	0.0008	19.6838	0.0508	18.6346
90	80.730	0.0124	1594.607	0.0006	19.7523	0.0506	18.8712
95	103.035	0.0097	2040.694	0.0005	19.8059	0.0505	19.0689
100	131.501	0.0076	2610.025	0.0004	19.8479	0.0504	19.2337

n	Single Payment Compound-Amount Factor (F/P,i,n)	Present-Worth Factor (P/F,i,n)	Equal-Payment Series Compound-Amount Factor (F/A,i,n)	Sinking-Fund Factor (A/F,i,n)	Present-Worth Factor (P/A,i,n)	Capital-Recovery Factor (A/P,i,n)	Uniform Gradient-Series Factor (A/G,i,n)
1	1.040	0.9615	1.000	1.0000	0.9615	1.0400	0.0000
2	1.082	0.9246	2.040	0.4902	1.8861	0.5302	0.4902
3	1.125	0.8890	3.122	0.3204	2.7751	0.3604	0.9739
4	1.170	0.8548	4.246	0.2355	3.6299	0.2755	1.4510
5	1.217	0.8219	5.416	0.1846	4.4518	0.2246	1.9216
6	1.265	0.7903	6.633	0.1508	5.2421	0.1908	2.3857
7	1.316	0.7599	7.898	0.1266	6.0021	0.1666	2.8433
8	1.369	0.7307	9.214	0.1085	6.7328	0.1485	3.2944
9	1.423	0.7026	10.583	0.0945	7.4353	0.1345	3.7391
10	1.480	0.6756	12.006	0.0833	8.1109	0.1233	4.1773
11	1.539	0.6496	13.486	0.0742	8.7605	0.1142	4.6090
12	1.601	0.6246	15.026	0.0666	9.3851	0.1066	5.0344
13	1.665	0.6006	16.627	0.0602	9.9857	0.1002	5.4533
14	1.732	0.5775	18.292	0.0547	10.5631	0.0947	5.8659
15	1.801	0.5553	20.024	0.0500	11.1184	0.0900	6.2721
16	1.873	0.5339	21.825	0.0458	11.6523	0.0858	6.6720
17	1.948	0.5134	23.698	0.0422	12.1657	0.0822	7.0656
18	2.026	0.4936	25.645	0.0390	12.6593	0.0790	7.4530
19	2.107	0.4747	27.671	0.0361	13.1339	0.0761	7.8342
20	2.191	0.4564	29.778	0.0336	13.5903	0.0736	8.2091
21	2.279	0.4388	31.969	0.0313	14.0292	0.0713	8.5780
22	2.370	0.4220	34.248	0.0292	14.4511	0.0692	8.9407
23	2.465	0.4057	36.618	0.0273	14.8569	0.0673	9.2973
24	2.563	0.3901	39.083	0.0256	15.2470	0.0656	9.6479
25	2.666	0.3751	41.646	0.0240	15.6221	0.0640	9.9925
26	2.772	0.3607	44.312	0.0226	15.9828	0.0626	10.3312
27	2.883	0.3468	47.084	0.0212	16.3296	0.0612	10.6640
28	2.999	0.3335	49.968	0.0200	16.6631	0.0600	10.9909
29	3.119	0.3207	52.966	0.0189	16.9837	0.0589	11.3121
30	3.243	0.3083	56.085	0.0178	17.2920	0.0578	11.6274
31	3.373	0.2965	59.328	0.0169	17.5885	0.0569	11.9371
32	3.508	0.2851	62.701	0.0160	17.8736	0.0560	12.2411
33	3.648	0.2741	66.210	0.0151	18.1477	0.0551	12.5396
34	3.794	0.2636	69.858	0.0143	18.4112	0.0543	12.8325
35	3.946	0.2534	73.652	0.0136	18.6646	0.0536	13.1199
40	4.801	0.2083	95.026	0.0105	19.7928	0.0505	14.4765
45	5.841	0.1712	121.029	0.0083	20.7200	0.0483	15.7047
50	7.107	0.1407	152.667	0.0066	21.4822	0.0466	16.8123
55	8.646	0.1157	191.159	0.0052	22.1086	0.0452	17.8070
60	10.520	0.0951	237.991	0.0042	22.6235	0.0442	18.6972
65	12.799	0.0781	294.968	0.0034	23.0467	0.0434	19.4909
70	15.572	0.0642	364.290	0.0028	23.3945	0.0428	20.1961
75	18.945	0.0528	448.631	0.0022	23.6804	0.0422	20.8206
80	23.050	0.0434	551.245	0.0018	23.9154	0.0418	21.3719
85	28.044	0.0357	676.090	0.0015	24.1085	0.0415	21.8569
90	34.119	0.0293	817.983	0.0012	24.2673	0.0412	22.2826
95	41.511	0.0241	1012.785	0.0010	24.3978	0.0410	22.6550
100	50.505	0.0198	1237.624	0.0008	24.5050	0.0408	22.9800

Table E.13 8% Interest Factors for Annual Compounding Interest

n	Single Payment: Compound-Amount Factor (To Find F Given P) F/P, i, n	Single Payment: Present-Worth Factor (To Find P Given F) P/F, i, n	Equal-Payment Series: Compound-Amount Factor (To Find F Given A) F/A, i, n	Equal-Payment Series: Sinking-Fund Factor (To Find A Given F) A/F, i, n	Equal-Payment Series: Present-Worth Factor (To Find P Given A) P/A, i, n	Equal-Payment Series: Capital-Recovery Factor (To Find A Given P) A/P, i, n	Uniform Gradient-Series Factor (To Find A Given G) A/G, i, n
1	1.080	0.9259	1.000	1.0000	0.9259	1.0800	0.0000
2	1.166	0.8573	2.080	0.4808	1.7833	0.5608	0.4808
3	1.260	0.7938	3.246	0.3080	2.5771	0.3880	0.9488
4	1.360	0.7350	4.506	0.2219	3.3121	0.3019	1.4040
5	1.469	0.6806	5.867	0.1705	3.9927	0.2505	1.8465
6	1.587	0.6302	7.336	0.1363	4.6229	0.2163	2.2764
7	1.714	0.5835	8.923	0.1121	5.2064	0.1921	2.6937
8	1.851	0.5403	10.637	0.0940	5.7466	0.1740	3.0985
9	1.999	0.5003	12.488	0.0801	6.2469	0.1601	3.4910
10	2.159	0.4632	14.487	0.0690	6.7101	0.1490	3.8713
11	2.332	0.4289	16.645	0.0601	7.1390	0.1401	4.2395
12	2.518	0.3971	18.977	0.0527	7.5361	0.1327	4.5958
13	2.720	0.3677	21.495	0.0465	7.9038	0.1265	4.9402
14	2.937	0.3405	24.215	0.0413	8.2442	0.1213	5.2731
15	3.172	0.3153	27.152	0.0368	8.5595	0.1168	5.5945
16	3.426	0.2919	30.324	0.0330	8.8514	0.1130	5.9046
17	3.700	0.2703	33.750	0.0296	9.1216	0.1096	6.2038
18	3.996	0.2503	37.450	0.0267	9.3719	0.1067	6.4920
19	4.316	0.2317	41.446	0.0241	9.6036	0.1041	6.7697
20	4.661	0.2146	45.762	0.0219	9.8182	0.1019	7.0370
21	5.034	0.1987	50.423	0.0198	10.0168	0.0998	7.2940
22	5.437	0.1840	55.457	0.0180	10.2008	0.0980	7.5412
23	5.871	0.1703	60.893	0.0164	10.3711	0.0964	7.7786
24	6.341	0.1577	66.765	0.0150	10.5288	0.0950	8.0066
25	6.848	0.1460	73.106	0.0137	10.6748	0.0937	8.2254
26	7.396	0.1352	79.954	0.0125	10.8100	0.0925	8.4352
27	7.988	0.1252	87.351	0.0115	10.9352	0.0915	8.6363
28	8.627	0.1159	95.339	0.0105	11.0511	0.0905	8.8289
29	9.317	0.1073	103.966	0.0096	11.1584	0.0896	9.0133
30	10.063	0.0994	113.283	0.0088	11.2578	0.0888	9.1897
31	10.868	0.0920	123.346	0.0081	11.3498	0.0881	9.3584
32	11.737	0.0852	134.214	0.0075	11.4350	0.0875	9.5197
33	12.626	0.0789	145.951	0.0069	11.5139	0.0869	9.6737
34	13.690	0.0731	158.627	0.0063	11.5869	0.0863	9.8208
35	14.785	0.0676	172.317	0.0058	11.6546	0.0858	9.9611
40	21.725	0.0460	259.057	0.0039	11.9246	0.0839	10.5699
45	31.920	0.0313	386.506	0.0026	12.1084	0.0826	11.0447
50	46.902	0.0213	573.770	0.0018	12.2335	0.0818	11.4107
55	68.914	0.0145	848.923	0.0012	12.3186	0.0812	11.6902
60	101.257	0.0099	1253.213	0.0008	12.3766	0.0808	11.9015
65	148.780	0.0067	1847.248	0.0006	12.4160	0.0806	12.0602
70	218.606	0.0046	2720.080	0.0004	12.4428	0.0804	12.1783
75	321.205	0.0031	4002.557	0.0003	12.4611	0.0803	12.2658
80	471.955	0.0021	5886.935	0.0002	12.4735	0.0802	12.3301
85	693.456	0.0015	8655.706	0.0001	12.4820	0.0801	12.3773
90	1018.915	0.0010	12723.939	0.0001	12.4877	0.0801	12.4116
95	1497.121	0.0007	18701.507	0.0001	12.4917	0.0801	12.4365

Table E.14 9% Interest Factors for Annual Compounding Interest

n	Single Payment: Compound-Amount Factor (To Find F Given P) F/P, i, n	Single Payment: Present-Worth Factor (To Find P Given F) P/F, i, n	Equal-Payment Series: Compound-Amount Factor (To Find F Given A) F/A, i, n	Equal-Payment Series: Sinking-Fund Factor (To Find A Given F) A/F, i, n	Equal-Payment Series: Present-Worth Factor (To Find P Given A) P/A, i, n	Equal-Payment Series: Capital-Recovery Factor (To Find A Given P) A/P, i, n	Uniform Gradient-Series Factor (To Find A Given G) A/G, i, n
1	1.090	0.9174	1.000	1.0000	0.9174	1.0900	0.0000
2	1.188	0.8417	2.090	0.4785	1.7591	0.5685	0.4785
3	1.295	0.7722	3.278	0.3051	2.5313	0.3951	0.9426
4	1.412	0.7084	4.573	0.2187	3.2397	0.3087	1.3925
5	1.539	0.6499	5.985	0.1671	3.8897	0.2571	1.8282
6	1.677	0.5963	7.523	0.1329	4.4859	0.2229	2.2498
7	1.828	0.5470	9.200	0.1087	5.0330	0.1987	2.6574
8	1.993	0.5019	11.028	0.0907	5.5348	0.1807	3.0512
9	2.172	0.4604	13.021	0.0768	5.9953	0.1668	3.4312
10	2.367	0.4224	15.193	0.0658	6.4177	0.1558	3.7978
11	2.580	0.3875	17.560	0.0570	6.8052	0.1470	4.1510
12	2.813	0.3555	20.141	0.0497	7.1607	0.1397	4.4910
13	3.066	0.3262	22.953	0.0436	7.4869	0.1336	4.8182
14	3.342	0.2993	26.019	0.0384	7.7862	0.1284	5.1326
15	3.642	0.2745	29.361	0.0341	8.0607	0.1241	5.4346
16	3.970	0.2519	33.003	0.0303	8.3126	0.1203	5.7245
17	4.328	0.2311	36.974	0.0271	8.5436	0.1171	6.0024
18	4.717	0.2120	41.301	0.0242	8.7556	0.1142	6.2687
19	5.142	0.1945	46.018	0.0217	8.9501	0.1117	6.5236
20	5.604	0.1784	51.160	0.0196	9.1286	0.1096	6.7675
21	6.109	0.1637	56.765	0.0176	9.2923	0.1076	7.0006
22	6.659	0.1502	62.873	0.0159	9.4424	0.1059	7.2232
23	7.258	0.1378	69.532	0.0144	9.5802	0.1044	7.4358
24	7.911	0.1264	76.790	0.0130	9.7066	0.1030	7.6384
25	8.623	0.1160	84.701	0.0118	9.8226	0.1018	7.8316
26	9.399	0.1064	93.324	0.0107	9.9290	0.1007	8.0156
27	10.245	0.0976	102.723	0.0097	10.0266	0.0997	8.1906
28	11.167	0.0896	112.968	0.0089	10.1161	0.0989	8.3572
29	12.172	0.0822	124.135	0.0081	10.1983	0.0981	8.5154
30	13.268	0.0754	136.308	0.0073	10.2737	0.0973	8.6657
31	14.462	0.0692	149.575	0.0067	10.3428	0.0967	8.8083
32	15.763	0.0634	164.037	0.0061	10.4063	0.0961	8.9436
33	17.182	0.0582	179.800	0.0056	10.4645	0.0956	9.0718
34	18.728	0.0534	196.982	0.0051	10.5178	0.0951	9.1933
35	20.414	0.0490	215.711	0.0046	10.5668	0.0946	9.3083
40	31.409	0.0318	337.882	0.0030	10.7574	0.0930	9.7957
45	48.327	0.0207	525.859	0.0019	10.8812	0.0919	10.1603
50	74.358	0.0135	815.084	0.0012	10.9617	0.0912	10.4295
55	114.408	0.0088	1260.092	0.0008	11.0140	0.0908	10.6261
60	176.031	0.0057	1944.79	0.0005	11.0480	0.0905	10.7683
65	270.846	0.0037	2998.28	0.0003	11.0701	0.0903	10.8702
70	416.730	0.0024	4619.22	0.0002	11.0845	0.0902	10.9427
75	641.191	0.0016	7113.23	0.0002	11.0938	0.0902	10.9940
80	986.552	0.0010	10950.57	0.0001	11.0999	0.0901	11.0299
85	1517.932	0.0007	16854.80	0.0001	11.1038	0.0901	11.0551
90	2335.527	0.0004	25939.18	0.0001	11.1064	0.0900	11.0726
95	3593.497	0.0003	39916.63	0.0000	11.1080	0.0900	11.0847

Bibliography/ References

The following books are references that support the text in this book. This list also includes resources for further information, by chapter topic.

Chapter 1

1. Binder, Stephen, *Corporate Facility Planning*. New York, NY: McGraw-Hill, Inc., 1989.

2. Burrus, Daniel, with Gittines, Roger, *Techno Trends*. New York, NY: Harper Business, 1993.

3. Covey, Stephen R., *The 7 Habits of Highly Effective People*. New York, NY: Simon & Schuster, 1989.

4. Davidow, William H. and Uttal, Bro, *Total Customer Service—The Ultimate Weapon*. New York, NY: Harper Perennial, 1989.

5. Delavigne, Kenneth T. and Robertson, J. Daniel, *Deming's Profound Changes* Englewood Cliffs, NJ: PTR Prentice Hall, 1994.

6. Drucker, Peter F., *Managing for the Future*. New York, NY: Penguin Group, 1992.

7. Hamer, Jeffrey, *Facility Management Systems* New York, NY: Van Nostrand Reinhold Company, Inc., 1988.

8. Imparato, Nicholas and Harari, Oren, *Jumping the Curve*. San Francisco, CA: Jossey-Bass, Inc., 1994.

9. Joiner, Brian L., *Fourth Generation Management*. New York, NY: McGraw-Hill, Inc., 1994.

10. Porter, Michael E., *Competitive Strategy*. New York, NY: The Free Press, 1980.

11. Thompson, Arthur A., Jr. and Strickland, A.J. III., *Strategy Formulation and Implementation—Tasks of the General Manager*, Third Edition. Plano, Texas: Business Publications, Inc., 1986.

12. Walton, Mary, *The Deming Management Method*. New York, NY: The Putnam Publishing Group, 1986.

Chapter 2

1. Davis, Gerald and Ventre, Francis T., Editors, *Performance of Buildings and Serviceability of Facilities*. Philadelphia, PA: American Society for Testing and Materials (ASTM), 1990.

2. Delavigne, Kenneth T. and Robertson, J. Daniel, *Deming's Profound Changes*. Englewood Cliffs, NJ: PTR Prentice Hall, 1994.

3. Kahn, Sanders A. and Case, Frederick E., *Real Estate Appraisal and Investment*. New York, NY: John Wiley & Sons, 1977.

4. Preiser, Wolfgang F.E.; Rabinowitz, Harvey Z.; and White, Edward ' *Post-Occupancy Evaluation*. New York, NY: Van Nostrand Reinhold Company, Inc., 1988.

5. Rush, Richard D., Editor, *The Building Systems Integration Handbook*. New York, NY: The American Institute of Architects and John Wile & Sons, 1986.

6. Vischer, Jacqueline C., *Environmental Quality in Offices* New York, NY: Van Nostrand Reinhold, 1989.

Chapter 3

1. American National Standard Practice for Industrial Lighting (RP-7-1991), Illuminating Engineering Society of North America, American National Standards Institute.

2. American National Standard Practice for Office Lighting (RP-1-1993), Illuminating Engineering Society of North America, American National Standards Institute.

3. American Society of Heating, Refrigerating and Air Conditioning Engineers, Inc. *ASHRAE Guideline 1-1989, Guideline for Commissionin of HVAC Systems* Atlanta, GA: 1989.

4. American Society of Heating, Refrigerating and Air Conditioning Engineers, Inc. *ASHRAE Guideline 55-1992, Thermal Environmental Conditions for Human Occupancy*, Atlanta, GA: 1992.

5. American Society of Heating, Refrigerating, and Air Conditioning Engineers, Inc. *ASHRAE Guideline 62-1989, Ventilation for Acceptable Indoor Air Quality*, Atlanta, GA: 1989.

6. Aronoff, Stan and Kaplan, Audrey, *Total Workplace Performance—Rethinking the Office Environment* Ottawa, Canada: WDL Publications, 1995.

7. Camp, Robert C., *Benchmarking—The Search For Industry Best Practices that Lead to Superior Performance*. Milwaukee, WI: Quality Press, 1989.

8. Cushman, Robert; Lipman, Andrew and Sugarman, Alan, *High Tech Real Estate.* Homewood, IL: Dow Jones-Irwin, 1985.

9. Delavigne, Kenneth T. and Robertson, J. Daniel, *Deming's Profound Changes.* Englewood Cliffs, NJ: PTR Prentice Hall, 1994.

10. Goldratt, Eliyahu M., *The Goal.* Hudson, NY: North River Press, Inc., 1984.

11. Harrington, James H., *Business Process Improvement* New York, NY: McGraw-Hill, Inc., 1991.

12. Hartmann, Edward H., *Successfully Installing TPM in a Non-Japanese Plant.* Pittsburgh, PA: TPM Press, 1992.

13. Ioannides, Stathis, "New Life Theory" (report). King of Prussia, PA: SKF USA.

14. International Facility Management Association, *Benchmarks III—Research Report # 18,* Houston, TX, 1997.

15. Joiner, Brian L., *Fourth Generation Management.* New York, NY: McGraw-Hill, Inc., 1994.

16. Juran, J.M. and Gryna, Frank M., Jr., *Quality Planning and Analysis—From Product Development Through Use*, Second Edition. New York, NY: McGraw-Hill Book Company, 1980.

17. Kaiser, Harvey H., *The Facilities Manager's Reference* Kingston, MA: R.S. Means Co., Inc., 1989.

18. Lueder, Rani, *The Ergonomics Payoff.* New York, NY: Nichols Publishing Company, 1986.

19. Maggard, Bill N., *TPM That Works, A Guide for Implementing TPM.* Pittsburgh, PA: TPM Press, 1992.

20. Rex O. Dillow, Editor-in-Chief, *Facilities Management—A Manual for Plant Administration,* Second Edition. Alexandria, VA: Association of Physical Plant Administrators of Universities and Colleges, 1989.

21. Ruck, N.C., *Building Design and Human Performance.* New York, NY: Van Nostrand Reinhold, 1989.

22. United States Environmental Protection Agency, *Guide to Pollution Prevention—The Commercial Printing Industry,* August, 1990.

23. Vischer, Jacqueline C., *Environmental Quality in Offices.* New York, NY: Van Nostrand Reinhold, 1989.

Chapter 4

1. Dell'Isola, Alphonse, *Value Engineering: Practical Applications* Kingston, MA: R.S. Means Co., Inc., 1997.

2. Fallon, Carlos, *Value Analysis,* Second Revised Edition. New York, NY: Wiley-Interscience, a division of John Wiley & Sons, 1978.

3. Kirk, Stephen J. and Spreckelmeyer, Kent F., *Creative Design Decisions—A Systematic Approach to Problem Solving in Architecture.* New York, NY: Van Nostrand Reinhold Company, 1988.

4. Macedo, Manuel C., Jr.; Dobrow, Paul V.; and O'Rourke, Joseph J., *Value Management for Construction.* New York, NY: John Wiley & Sons, 1978.

5. Nadler, Gerald and Hibino, Shozo, *Breakthrough Thinking*. Rucklin, CA: Prima Publishing and Communications, 1990.

6. O'Brien, James J., *Value Analysis in Design and Construction*. New York, NY: McGraw-Hill Book Company, 1976.

Chapter 5

1. Adams, John R., *Roles and Responsibilities of the Project Manager*. Drexel, PA: Project Management Institute, 1982.

2. Burke, Rory, *Project Management Planning and Control*, 2nd Edition. New York, NY: John Wiley & Sons, 1992.

3. Cable, Dwayne, *Organizing for Project Management*. Drexel, PA: Proje Management Institute, 1982.

4. Cleland, David I. and King, William R., *Systems Analysis and Project Management*. New York, NY: McGraw-Hill Book Company, 1983.

5. The Construction Industry Institute, *Work Packaging for Project Control*. Austin, TX: University of Texas at Austin, Construction Industry Institute, 1988.

6. Galbraith, Jay R., *Organizing for the Future*. San Francisco, CA: Jossey-Bass, Inc., 1993.

7. Kerzner, Harold, *Project Management—A Systems Approach to Planning, Scheduling and Controlling*, Second Edition. New York, NY: Van Nostrand Reinhold Company, 1984.

8. Kirchof, Nicki S., *Conflict Management for Project Managers*. Drexel, P/ Project Management Institute, 1989.

9. Leavitt, Jeffrey S., *Total Quality Through Project Management*. New York, NY: McGraw-Hill, Inc., 1994.

10. Martin, Martin D., *Contract Administration for the Project Manager*. Drexel Hill, PA: Project Management Institute, 1987.

11. Meredith, Jack R. and Mantel, Samuel J., Jr., *Project Management—A Managerial Approach* New York, NY: John Wiley & Sons, 1985.

12. O'Connor, J.T. and Vickroy, *Control of Construction Project Scope—A Report to the Construction Industry Institute*. Austin, TX: University of Texas at Austin, Construction Industry Institute, 1986.

13. Thompson, Charles ("Chic"), *What a Great Idea!* New York, NY: Harper Collins Publishers, 1992.

14. von Oech, Roger, *A Whack on the Side of the Head*. New York, NY: Warner Books, Inc., 1983.

15. von Oech, Roger, *A Kick in the Seat of the Pants*. New York, NY: Harper & Row Publishers, Inc., 1986.

16. Pena, William, *Problem Seeking—An Architectural Programming Prime* Third Edition. Washington, DC: AIA Press, 1987.

17. Wycoff, Joyce, *Mindmapping—Your Personal Guide to Exploring Creativity and Problem-Solving*. New York, NY: Berkley Publishing Group, 1991.

Chapter 6

1. Dell'Isola, Alphonse J. and Kirk, Stephen J., *Life Cycle Costing for Design Professionals*. New York, NY: McGraw-Hill Book Company, 1981.

2. Ferry, Douglas J. and Brandon, Peter S., *Cost Planning of Buildings* New York, NY: Granada Publishing, Ltd., 1984.

3. Humphreys, Kenneth K., Editor, *Project and Cost Engineers' Handbook*. New York, NY: Marcel Dekker, Inc., 1984.

4. Jelen, Frederic C. and Black, James H., *Cost and Optimization Engineering*. New York, NY: McGraw-Hill Book Company, 1983.

5. Killingsworth, Roger, *Cost Control in Building Design*. Kingston, MA: R.S. Means Company, Inc., 1988.

6. Mansfield, G.L., *Bidding and Estimating Procedures for Construction*. Reston, VA: Reston Publishing Company, Inc., 1983.

7. Messner, Stephen D.; Schreiber, Irving; Lyon, Victor L.; and Ward, Robert L., *Marketing Investment Real Estate—Finance Taxation Techniques*, Second Edition. Chicago, IL: Realtors National Marketing Institute of the National Association of Realtors, 1982.

8. O'Connor, J.T. and Vickroy, *Control of Construction Project Scope—A Report to the Construction Industry Institute*, Austin, TX: University of Texas at Austin, Construction Industry Institute, 1986.

9. Parker, Donald E. and Dell'Isola, Alphonse, *Project Budgeting for Buildings*. New York, NY: Van Nostrand Reinhold, 1991.

10. Palmer, William J.; Coombs, William E.; and Smith, Mark A., *Construction Accounting and Financial Management*, Fifth Edition. New York, NY: McGraw-Hill, Inc., 1995.

11. Stewart, Rodney D. and Wyskida, Richard M., *Cost Estimator's Reference Manual*. New York, NY: John Wiley & Sons, 1987.

Chapter 7

1. Gido, Jack, *An Introduction to Project Planning*, Second Edition. New York, NY: Industrial Press, Inc., 1985.

2. Harris, Robert B., *Precedence and Arrow Networking Techniques for Construction*. New York, NY: John Wiley & Sons, 1978.

3. Horsley, F. William, *Means Scheduling Manual*, Third Edition. R.S. Means Company, Inc., Kingston, MA, 1991.

4. Kerzner, Harold, *Project Management—A Systems Approach to Planning, Scheduling and Controlling*, Second Edition. New York, NY: Van Nostrand Reinhold Company, 1984.

5. Levine, Harvey A., *Project Management—Using Microcomputers*. Berkeley, CA: Osborne McGraw Hill, 1986.

6. O'Brien, James J., *CPM in Construction Management*. New York, NY: McGraw-Hill Book Company, 1971.

7. Pierce, David R., Jr., *Project Planning and Control for Construction* Kingston, MA: R.S. Means Company, Inc., 1988.

8. Stevens, James D., *Techniques for Construction Network Scheduling*, New York, NY: McGraw-Hill, 1990.

Chapter 8

1. Adrian, James J., *CM: The Construction Management Process*. Reston, VA: Reston Publishing Company, Inc., 1981.
2. The American Institute of Architects, *The Architect's Handbook of Professional Practice* Washington, D.C.: The American Institute of Architects, 1988.
3. American Society of Civil Engineers, *Quality in the Constructed Project: A Guideline for Owners, Designers, and Constructors* (Manual o Professional Practice, Volume 1, Preliminary Edition of Trial Use and Comment.) New York, NY: American Society of Civil Engineers, 1988.
4. Cavendish, Penny and Martin, Martin D., *Negotiating & Contracting f Project Management*. Drexel Hill, PA: Project Management Institute, 1982.
5. The Construction Specifications Institute, *Manual of Practice*. Alexandria, VA: The Construction Specifications Institute, 1996.
6. Cushman, Robert F. and Palmer, William J., *Businessman's Guide to Construction*. Princeton, NJ: Dow Jones Books, 1980.
7. Gilbreath, Robert D., *Managing Construction Contracts* New York, NY: John Wiley & Sons, 1983.
8. Halpin, Daniel and Woodhead, Ronald W., *Construction Management* New York, NY: John Wiley & Sons, 1980.
9. Levin, Paul, *Claims and Changes: Handbook for Construction Contract Management*. Silver Spring, MD: Construction Industry Press, 1981
10. Martin, Martin D.; Teagarden, C. Claude; and Lambreth, Charles F. *Contract Administration for the Project Manager*. Drexel Hill, PA: Project Management Institute, 1987.
11. Papageorge, Thomas E., *Risk Management for Building Professionals*. Kingston, MA: R.S. Means Company, Inc., 1988.

Chapter 9

1. American Society of Civil Engineers, *Quality in the Constructed Project: A Guideline for Owners, Designers and Constructors* (Manual o Professional Practice, Volume 1, Preliminary Edition of Trial Use and Comment.) New York, NY: American Society of Civil Engineers, 1988.
2. Dinsmore, Paul C., *Human Factors in Project Management*. New York, NY: American Management Association, 1984.
3. Dinsmore, Paul C., *The Project Manager's Work Environment: Coping with Time and Stress*. Drexel Hill, PA: Project Management Institute, 1988.

4. Grimes, J. Edward, *Construction Paperwork An Efficient Management System* Kingston, MA: R.S. Means Company, Inc., 1989.
5. Stuckenbruck, Linn C., *Team Building for Project Managers.* Drexel Hill, PA: Project Management Institute, 1988.

Chapter 10

1. American Society of Civil Engineers, *Quality in the Constructed Project: A Guideline for Owners, Designers and Constructors* (Manual of Professional Practice, Volume 1, Preliminary Edition of Trial Use and Comment.) New York, NY: American Society of Civil Engineers, 1988.
2. Belasco, James A. and Stayer, Ralph C., *Flight of the Buffalo.* New York, NY: Warner Books, Inc., 1993.
3. Dannithorne, L., *The West Point Way of Leadership.* New York, NY: Bantam-Doubleday-Dell, 1993.
4. DePree, Max, *Leadership Jazz* New York, NY: Dell Publishing, 1992.
5. Harrington-Mackin, Deborah, *The Team Building Tool Kit* New York, NY: American Management Association, 1994.
6. Jablonski, Joseph R., *Implementing TQM* Albuquerque, NM: Technical Management Consortium, Inc., 1994.
7. Kolbe, Kathy, *The Conative Connection—Uncovering the Link Between Who You Are and How You Perform.* Reading, MA: Addison-Wesley Publishing Company, Inc., 1990.
8. Larson, Carl E. and LaFasto, Frank M., *Team Work.* Newbury Park, CA: Sage Publications, 1989.
9. Peters, Tom, *Liberation Management—Necessary Disorganization for the Nanosecond Nineties.* New York, NY: Alfred A. Knopf, 1992.
10. Lundy, James, Ph.D., *Lead, Follow, or Get Out of The Way.* San Diego, CA: Pfeiffer & Company, 1993.
11. Phillips, Donald T., *Lincoln on Leadership* New York, NY: Warner Books, 1992.
12. Portnoy, Robert A., *Leadership! What Every Leader Should Know About People.* Englewood Cliffs, NJ: Prentice-Hall, Inc., 1986.
13. Stuckenbruck, Linn C., *Team Building for Project Managers.* Drexel Hill, PA: Project Management Institute, 1988.
14. Tichy, Noel M. and Sherman, Stratford, *Control Your Destiny or Someone Else Will.* New York, NY: Doubleday, 1993.

Index

A

accountability, 9, 124
Activity-On-Arrow (AOA) approach, 167, 169
Activity-On-Node (AON) approach, 167, 169, 172
activity relationships, 50, 54
adjacency relationships charts, 50, 54
affinity diagrams, 223, 225
agency construction management, 187
air quality, indoor
    benchmarks, 76–79
    building materials pollution, 77–78
    definition of, 77
    effects of building changes on, 77
    improving, 7
    maintenance and, 71
    ratios, 77
American Institute of Architects (AIA), 191
American Society for Testing & Materials (ASTM), 24
American Society of Heating, Refrigerating and Air-Conditioning Engineers (ASHRAE), 24, 40
    air quality standards, 76–77
    thermal quality standards, 79
Americans with Disabilities Act (ADA), 25, 81
architectural components, 32
Associated General Contractors (AGC), 191

Association for Facilities Engineers (AFE), 24
attitude surveys, customer, 14–15
    sample, 16–19
audits, energy, 50, 56, 57
authority, 126, 127

B

backlog reports, 65–66
backward pass, calculating, 171, 174
bar charts, 161–163
    advantages/disadvantages of, 163
behavioral building evaluations, 31
benchmarking, 39–83
    best affordable technology/ methodology and, 40
    data for, 40
    definition of, 39
    and energy costs, 50, 56–59
    environmental health/safety, 76–82
    objectives, 41
    occupant survey method, 44, 45
    for outsourcing, 110
    overall equipment effectiveness, 73, 74, 75
    partners, 40
    post-occupancy evaluations and, 30
    process in, 46–59
    ratio analysis method, 44
    sample study, 83
    selecting indicators for, 39–40
    and space utilization, 46–50, 51–55

and statistical process
control, 41–43
studies, 40–41
value engineering studies, 44, 46
and waste minimization, 59–76
benefits/risk analysis, 98
bid bonds, 183
bids
comparing, 195, 196–198
documents for, 194
forms for, 195, 196–198
in general contractor
method, 182–184
importance of detailed information
for, 184
invitation to bid, 194–199
open-book basis for, 185
bonds
bid, 183
construction payment, 193
construction performance, 193
performance payment, 183
brainstorming, 92, 95
and organizational structure,
120–121
and work breakdown structures,
131
Brill, Michael, 23–24
Broad Form comprehensive
insurance, 201–202
bubble diagrams, 50, 55
budget creep, 145–146
budgets
checklist for project, 146–147
cost information sources
for, 147–148
developing realistic, 141, 143–146
pre-acquisition evaluations and, 27
and work breakdown
structures, 131
Buffalo Organization for Social and
Technological Innovation
(BOSTI), 23–24
Building Owners and Managers
Association (BOMA) International, 25
Bureau of Labor Statistics, 81
business growth, 8

C

carbon dioxide levels, 78
cash flow elements, 151–152
cause and effect diagrams, 69, 223, 22
certificates, 193
insurance coverage, 201
change orders, 218
charts
adjacency relationships, 50, 54
bar, 161–163
decision, 113, 114
milestone, 160–161
organizational, 113, 116
process flow, 47, 50
chlorofluorocarbons (CFCs), 7–8
code compliance, 24
systems analysis for, 31–32
communication, 209–220
of change orders, 218
definition of effective, 209–210
and documentation, 213–217
and fear, 210
feedback in, 210
in functional organizations, 113,
115–116
and leadership, 227–228
and listening, 210–211
in meetings, 213
on-site, 218–219
and outsourcing, 111
on projects, 211–212
and schedules, 176
technologies, 10, 219–220
written, 212–213
compensation methods, 202–204
and general contractor method, 1
competition and cost control, 9–10
compounding interest, 247-248
computer-aided building
management, 219–220
computer-aided facility management
(CAFM), 219
computer technology, 10
and scheduling, 157
conceptual estimates, 146
conflict
in project life cycle, 124
project management, 126–128

resolving with decision charts, 113, 114
construction documents/drawings, 146
  and bid accuracy, 149
  communication about, 212–213
construction management method, 186–191
  agency, 187
  contractor, 188–190
  and guaranteed maximum price contracts, 191
  responsibility matrix for, 192
construction payment bonds, 193
construction performance bonds, 193
construction specifications, 193–194
Construction Specifications Institute, 147, 189
consultants, selecting, 221–222
contingencies, 147–148, 189
contracting/procurement methods, 181–205
  and compensation methods, 202–204
  construction management, 186–191
  and contract documents, 191, 193–194
  creative financing/contracting, 204–205
  design/build, 184–186
  general contractor, 181–184
  and insurance, 201–202
  invitation to bid, 194–199
  and job site safety programs, 200–201
  and owner liabilities, 199–201
contractors
  budget input from, 144–145
  general contractor method and, 181–184
  insurance for, 201–202
  references, 221–222
  selecting for project teams, 221–222
  value engineering change proposals, 99
  "Contractor's Qualification Statement," 221
contracts
  cost-plus-fixed fee, 191

  documents, 191, 193–194
  filing, 213–214
  general/supplementary conditions of, 195, 199
  guaranteed maximum price, 191
  lump sum, 191
  maintenance, 69–71
  milestones in, 160–161
  program requirement clauses, 99
  scope documentation for, 133–134
  and scope of work, 133
  standard forms for, 191, 193
  value engineering incentive clauses, 99
  "Control of Construction Project Scope, A Report for the Construction Industry Institute" (O'Connor, Vickroy), 133
core competencies, 15
corporate culture/environment, 9, 15, 20
cost/benefit analysis. See project management
cost management, 141–157
  budget checklist for, 146–147
  cash flow elements, 151–152
  cost information sources, 147–148
  cost trade-off studies, 148–149
  in different phases of projects, 142
  estimating project costs, 143–146
  financial evaluation techniques, 149–151
  life cycle cost analysis, 151–157
  monitoring/reporting costs, 148
  time value of money and, 152–153
  tools for, 141
  and value engineering, 148
cost plus fixed fee fee structure, 191, 203–204
cost plus incentive fee structure, 203
cost plus percent of cost fee structure, 204
costs
  accountability and, 9
  categories in Life Cycle Cost Analysis, 151–152
  controlling, 9–10
  converting to equivalent annual charge, 156–157

converting to present worth, 153–154

data sources, 90

and design to budget, 99

energy, 50, 56–59

estimating project,143–146

expense ratios, 23–24

first vs operating/maintenance, 102

life cycle, 99, 111, 141, 149–151

maintenance, 60, 62–76

overruns and scope management, 129–130

sales/production rates and, 141

sample conceptual estimate, 243–246

and scheduling, 159

technology and, 10

trade-off studies of, 148–149

types of, 143

cost-to-worth ratio, 88

creativity, 95, 98

and functional organizations, 115

and positive/negative forces analysis, 226

critical path method (CPM), 165–175. *See also* project management

activities in, 167

Activity-On-Arrow method, 167, 169

Activity-On-Node method, 167,169

advantages/disadvantages of, 175

backward pass calculation, 171, 174

determining critical path, 165, 167–171

float calculation, 173, 174

forward pass calculation, 171

and resource leveling, 165, 175, 177–179

sample schedule, 166

time unit selection for, 167, 171

Customer Attitude Survey Rating Sheets, 15, 16–17, 227

customers

attitude survey, 14–15, 16–17, 227

communicating with, 211

complaints, 15, 18–19

external, 13

internal, 13

knowing, 13–15

and organization structure, 120

in project planning, 129

reevaluating requirements of, 87–88

understanding expectations of, 12

customer service, 121

D

damages, 199

data

benchmarking, 40

for contract negotiations, 133–134

manufacturers', 134

for process improvement, 5

sources of cost, 90

statistical process control, 41–42

decision charts, 113, 114

decision-making

with affinity diagrams, 223, 225

with cause and effect diagrams, 223, 224

with FAST diagrams, 222–223

and information quality, 23

with positive/negative forces analysis, 226

techniques for, 222–226

deliverables

definition of, 128

in scope statements, 130

dependencies, 167, 168

deployment flow charts, 50, 52

deregulation, utility, 58–59

design

to budget, 99

and construction specifications, 193–194

to cost, 149

development of, 146

evaluating alternative, 148–149

phases in, 146

programming and, 26–27

schematic, 146

design/build method, 184–186

advantages/disadvantages of, 185–186

diagnostic surveys, 25–26

direct costs, 143

Discounted Payback Period (DPP), 153. *See also* Life Cycle Cost Analysis

discount rate, 152–153
documentation
    for cost estimates, 145
    photographs as, 214
    project scope, 133–134
    record filing systems for, 213–217
duct distribution systems, 56
Dupont Corporation, 165

E

early finish (EF) time, 171
early start (ES) time, 171
earthwork, 204
electrical analysis, 63
electrical systems evaluation, 31–32
e-mail, 219
energy conservation. *See also* waste minimization
    and creative financing, 204–205
energy costs, 50, 56–59
    performance ratios, 56, 58
    utility deregulation and, 58–59
energy system analysis, 34
engineering economics. *See* Time Value of Money
Engineer's Joint Contract Documents Committee (EJCDC), 191
environment, corporate. *See* corporate culture/environment
environmental evaluations, 32
Environmental Protection Agency (EPA), 6
    Waste Minimization Opportunity Assessments, 59
equipment. *See also* maintenance costs, 24
    creative financing for, 204–205
    inventory sheets, 134, 136–137
    overall effectiveness, 73, 74, 75
    and space utilization, 47
Equivalent Annual Charges (EAC), 153
See also Life Cycle Cost Analysis
    converting to, 156–157
ergonomics, 82
essential functions, identifying, 88
estimates
    for change orders, 218
    contingency factor in, 147–148
    cost of, 143–145

importance of specific information for, 144–145
    schematic/conceptual, 146
    updating, 145–146
    and work breakdown structures, 131
ethics, 227
evaluation. *See* performance evaluations, facility
executives
    and communication, 10–11, 211–212
    gaining project support from, 149–151
    measurements used by, 12–13
expense ratios, 23–24
external customers, 13. *See also* customers

F

facilities as cost centers vs business resources, 6, 10
facilities programming. *See* programming, facilities
facility management. *See also* Total Productive Facilities Management (TPFM)
    automation of, 219–220
    current challenges in, 3–4, 6–8
    customer-oriented, 111–112
    definition of, 5
    facility performance evaluations, 23–37
    history of, 6
    importance of, 3–4
    performance categories, 46
    performance evaluations, 23–37
    project-driven, 108, 111–112
    seven tensions of, 7
    value engineering applications in, 86–89
facility managers
    continuous improvement for, 11–13
    current challenges for, 6–7
    management self-evaluation for, 11
    skills for, 10–11
FAST (Function Analysis System Technique) diagrams, 91–95
    advantages of, 91

for determining customer requirements, 129
sample, 93, 94, 96
summary of, 97
task/customer-oriented, 91, 93, 94
for team problem-solving, 222–223
Faults and Complaints Rating Sheets, 15, 18, 19
faxes, 219
feedback, 13
fee structures, 202–204
filing systems, 213–217
financial evaluation techniques, 149–151
financial reserve studies, 25
financing
    costs of, 152
    creative, 204–205
fire protection, 32
fishbone diagrams, 223, 224
fixed costs, 143
fixed price plus incentive fee structure, 203
flexibility
    in building space, 8
    HVAC, 79
float, 171. *See also* critical path method (CPM)
    calculating total/free, 173, 174
    definition of, 165
flow charts
    and bubble diagrams, 50, 55
    deployment, 50, 52
    network scheduling, 163–164
    process, 47, 50
    on waste-generating processes, 60
focus groups, 13
forward pass, calculating, 171
free float (FF), 174
functional building evaluations, 30–32
functional organizations, 113, 115–116
    problems in, 115–116
function analysis. *See* value engineering functions, defining, 90–91
furniture costs, 24

G

Gantt, Henry, 161
Gantt Charts, 161–163

general conditions, 195, 199
general construction management, 188–190, 192
general contractor method, 181–184
government regulation, 6, 7
groupware, 219
growth
    business, 8
    and space utilization, 47
guaranteed maximum price (GMP) contracts, 191

H

hazardous waste evaluations, 32
health benchmarks, 76–82
higher order functions, 91
hoteling, 10
humidity control, 79
HVAC systems
    and air quality, 7
    and chlorofluorocarbons, 7–8
    and energy conservation, 56–59
    manufacturers' data on, 134
    post-occupancy evaluation of, 28
    technical evaluation of, 31–32, 36–37

I

Illuminating Engineers Society (IES), 25, 80
illumination quality, 80
improvements, prioritizing, 23–24
indirect costs, 143
initial capital investments, 152
insurance, 200, 201–202
interest, compounding, 247-248
internal customers, 13. *See also* customers
International Facility Management Association (IFMA), 5, 25, 40
Internet, the, 219
interviews
    for customer expectations, 13
    for organizational structure design, 120
invitation to bid, 194–199
    bid bond, 195
    bidders instructions in, 194
    bid form, 195, 196–198

contract conditions, 195,199
    information for bidders in, 195

J

janitorial activities, 71
Japan, value engineering in, 102
job plans, value engineering, 89–99
    creativity phase, 95, 98
    definition of, 89
    development phase, 98–99
    evaluation phase, 98
    and FAST diagrams, 91–95, 96–97
    functions in, 90–91
    implementation phase, 99, 100–101
    information phase, 90–95
    presentation phase, 99
job reports, 65
job sites
    communication on, 218–219
    safety programs, 200–201
Just-in-Time manufacturing, 175–176

K

Konar, Ellen, 23–24

L

labor utilization reports, 66, 67
late finish date (LFD), 171,174
late start date (LSD), 171
leadership, team, 227–229
lean manufacturing, 35
legal considerations, 10
    code compliance, 24
    and contracts, 193
    environmental/safety programs, 8
    indoor air quality, 78
    insurance, 201–202
    owner liabilities, 199–201
letters
    and liability, 200–201
    transmittal, 214, 217
liability, owner, 199–201
Life Cycle Cost Analysis, 99, 151–157
    cash flow elements, 151–152
    converting to present worth in, 153–154
    inflation factor in, 155–156
    for investment decisions, 8

results/output samples, 153–157
    steps in, 151
    and time value of money, 152–153
life cycle costs
    categories of, 141
    and financial evaluation techniques, 149–151
    and maintenance, 111
life safety laws, 32
lighting quality, 80
links, in CPM, 157
listening, 210–211
lump sum fee structure, 191, 202–203

M

maintenance. *See also* statistical
process control
    breakdown/emergency, 65
    checklist for, 68
    contracts, 69–71
    cost categories, 66
    costs, 60, 62–76
    cost vs replacement cost, 75
    determining root causes in, 69–71
    emphasis on, 6
    and energy management, 72
    expense ratio, 24
    and indoor air quality, 71
    predictive, 42, 63, 72
    preventive, 63, 65
    proactive, 42, 62–63
    ratios for performance
    measurement, 66, 69, 70, 71
    reliability-centered, 73, 76
    reports, 65–66
    staffing, 65–66
    strategies comparison, 63, 64
    total productive, 72–73
management approach
    continuous improvement in, 11–13
    evaluating, 11
Margulis, Stephen, 23–24
mass balance equation, 59–60, 61
MasterFormat, 147, 189
Material Safety Data Sheets, 78
matrix organizations, 115–116, 117, 120
    organizational chart, 119

Means Building Construction Cost Data, 147

Means Facilities Construction Cost Data, 147

Means Facilities Maintenance & Repair Cost Data, 25

mechanical systems evaluations, 31–32

meetings, 213, 214

Miles, Larry, 102

milestones
    definition of, 128
        milestone charts, 160–161
        in project planning, 128

mission, corporate, 12, 15

multidisciplinary approach. *See also* value engineering
    to building-related problems, 79
    in project management, 107
    in TPFM, 4–5

N

Nakajima, Seiichi, 72

network scheduling, 163–164, 169. *See also* critical path method (CPM)

noise, 31, 63, 81

nonessential functions, identifying, 88

Null, Robert, 14

O

objectives, project, 130

occupant surveys, 44, 45

Occupational Safety & Health Administration (OSHA), 6, 24
    and indoor air quality, 78
    and noise levels, 81

off-gassing, 77–78

oil analysis, 63

open-book basis, 185

opportunity costs, 143

organizational charts, 113

organizational structures, 112–120
    choosing, 117
    functional, 113, 115–116
    hybrid, 120–124, 125
    matrix, 117, 119–120
    pure project, 116–117, 118

out-gassing, 77–78

outsourcing, 9, 108–112
    definition of, 108

determining when to, 109–110
    maintenance, 69–71
    and value engineering studies, 102

out-tasking, 108–109. *See also* outsourcing

overall equipment effectiveness, 73, 74, 75

overtime reports, 66

P

Pareto's Law of Distribution, 88, 109–110

performance contracting, 204–205

performance criteria, selecting, 41

performance evaluations, facility, 23–37
    behavioral building, 31
    case histories, 32–37
    and code compliance, 24
    environmental, 32
    and facilities programming, 26–27
    humane working environments, 24–25
    post-occupancy, 27–28, 30
    pre-acquisition, 27, 28, 29, 30
    space requirement, 32
    standards for, 23–25
    technical building, 31–32
    types of, 25–32

performance improvement, 5

performance payment bonds, 183

performance requirements
See functions

permissible exposure levels (PELs),

personnel
    and air quality, 7
    and communication, 211–212
    costs, 24
    development, 121
    humane working environment for, 24–25
    influence of buildings on, 31
    involving, 218–219
    in matrix organizations, 117, 120
    and outsourcing, 110–111
    staffing levels, 108
    and work-related injuries, 81

photographs, 214

planning projects, 128–130

pollution
    from building materials, 77–78
    control equipment, 7
positive/negative forces analysis, 226
post-occupancy evaluations,
27–28, 30
pre-acquisition evaluations, 25, 27
    samples, 28, 29, 30
predecessor activities, 167
presentations, value engineering
study, 99
Present Worth of Expenditures
(PWE), 153, 247-248.
See also Life Cycle Cost Analysis
    converting to, 153–154
    tables, 247
problem-solving techniques, 222–226
process analysis/improvement, 5, 9
    continuous improvement in, 11–13
process flow charts, 47, 50
procurement methods. *See*
contracting/procurement
productivity, 10, 11
    and air quality, 7
    benchmarking and, 39–83
    and fear, 127
    and office design, 23–24
    and resource control, 159
programming, facilities, 26–27
    definition of, 26
    in planning process, 146
    worksheet for, 138
program requirement clauses, 99
project delivery methods, 181
project directories, 212
projectized organizations
See pure project organizations
project management, 107–139
    and communication, 211–212
    and conflict, 124, 126–128
    definition of, 107
    evaluating, 82
    for facility management, 108
    and organization structures,
    112–124
    and outsourcing/out-tasking,
    108–112
    planning/scoping projects in,
    128–130

    and project characteristics, 108
    and project life cycles, 124
    project manager role in, 126–128
    scope of work, communicating, 131,
    133–134
    success in, 112
    work breakdown structure,
    130–131, 132
project managers
    communication of scope of work,
    131, 133, 134
    developing project support, 149
    role of, 126–128
project objectives, 130
project orientation, 213
projects, definition of, 108
project statements, 130
pure project organizations, 116–117
    organization chart, 118

Q

quality
    control, 159
    customer expectations and, 13

R

ratio analysis method, 44
ratios
    analysis method, 44
    cost-to-worth, 88
    energy performance, 56, 58
    expense, 23–24
    indoor air quality, 77
    for maintenance performance
    measurement, 66, 69, 70, 71
    project performance, 82
    space utilization, 48
    waste minimization, 61, 62
records, filing systems for, 213–217
references, contractor, 221–222
reliability-centered maintenance, 73, 76
reliable processes, 42. *See also*
statistical process control
requests for authorization, 139
resource leveling, 175
    definition of, 165
    samples of, 177–179
resource utilization in matrix
organizations, 117, 120

responsibility, 126
retainages, 199
Return on Investment (ROI),153.
See also Life Cycle Cost Analysis
"Right to Know" Law, 78
risk
    analysis, 98
    and compensation methods,
    202–204
    and contracts, 191,193–194
    identifying project, 131
    and outsourcing, 110
    and project management,107
roofing systems, 31–32

S

safety, job site, 200–201
salvage value, 152
S.A.V.E. International, 89
scheduling, 159–179
    and bar charts, 161–163
    critical path method, 165–175
    master project, 189
    methods of, 159–160
    and milestone charts, 160–161
    network, 163–164
    objective of, 159
    and resource leveling, 175, 177–179
schematics
    design, 146
    and maintenance costs, 69, 71
scope
    communicating, 131, 133–134
    and contract negotiations, 133
    documentation, 133–134
    management, 129–130
    statements, 130
shop drawings, 212–213
sick-building syndrome, 79
Simple Payback technique, 149–150
slack. *See* float
soil tests, 31
space requirements evaluations, 32
space utilization
    analysis, 49
    and benchmarking, 46–50, 51–55
    definition of, 47
    for equipment operators, 134

evaluation of planning
    requirements, 32
    ratios, 48
    survey, 33
specifications
    construction, 193–194
    documentation of equipment,
    134, 138
    performance-oriented, 193–194
    project objectives and, 130
spreadsheets
    for customer needs, 120–121
    and Time Value of Money
    conversions, 151
staffing levels, 108
    maintenance, 65–66
stakeholders, 128
standards, facility performance, 23–2
statistical process control, 41–43
*Strategy Formulation and
Implementation* (Thompson,
Strickland), 15, 20
strengths/weaknesses analysis,
228-230
stress, and project management,
126–128
Strickland, A. J., III, 15, 20
structural engineers, 31
structural systems, evaluation of, 31
subcontractors, 181–184
successor activities, 167
sunk costs, 143
supplementary conditions, 195, 199
suppliers, working with,12
supporting/basic functions, 91
support systems, flexible, 8
SWOT (Strengths, Weaknesses,
Opportunities, Threats) analysis,
15, 20
    sample, 21–22

T

tasks, 91 *See also* functions
Taylor, Frederick, 161
team member awareness form,
228–230
teams, 221–229
    assembling, 221–222
    building, 127, 228
    characteristics of leaders, 227–22

and communication, 221
in construction management, 187–191
continuous improvement, 35
creativity and, 95, 98
decision-making/problem-solving for, 222–226
and FAST diagrams, 129
in matrix organizations, 117, 120
number of members for, 89
post-mortem evaluations of, 226–227
project manager role in, 126–128
in pure project organizations, 116
spaces for, 10
in TPFM, 4–5
value study, 89
team spaces, 10
technical building evaluations, 31–32
technology
best affordable, 40
communication, 219–220
and maintenance, 73
telecommuting, 10, 47
thermal load analysis, 31–32
thermal quality, 79–80
individual temperature control and, 78
thermography, 63
Thompson, Arthur A., Jr., 15, 20
Time Value of Money, 152–153
tables, 151, 156
total float (TF), 174
Total Productive Facilities Management (TPFM), 4–5
total productive maintenance (TPM), 72–73
transmittal letters, 214, 217
turning points, project, 128

U

ultrasonic testing, 63
undiscounted payback, 149–150
Uniformat System, 189
unit price fee structure, 204
Using Office Design to Increase Productivity, 23–24
utilities
deregulation of, 58–59
safety and, 201

V

value analysis. *See* value engineering
Value Analysis Systems Training (Null), 14
value engineering, 85–103. *See also* project management
change proposals, 99
as cost management tool, 148
definition of, 86
and design to budget, 99
for determining customer requirements, 129
facility management applications of, 86–89
and government agencies, 102
history/current use of, 102–103
incentive clauses, 99
for investment decisions, 8
job plans, 89–99
and life cycle costing, 99
sample analysis reports, 100–101, 238–242
selecting topics for, 88
studies and benchmarking, 44,46
value index, 88
value management. *See* value engineering
value mismatches, 95
variable costs, 143
variance analysis
and change orders, 218
of costs, 148
ventilation, symptoms of inadequate, 78
vibration analysis, 63
video conferences, 10, 219
virtual offices, 10

W

walk-through surveys, 25
waste, evaluations of hazardous, 32
waste minimization, 59–76
maintenance costs and, 60, 62–76
mass balance equation for, 59–60, 61
ratios, 61, 62
recycling and, 60
water treatment, 71

weaknesses/strengths analysis, 228–230

work breakdown structure, 130–131

    and cost estimating, 143

    errors/omissions in, 143

    organizing, 131

    sample, 132

    and scheduling, 159

workflow diagrams, 50, 53–54

work packages, 148

work plan outlines, 139

work restrictions, 195, 199

workshops, team, 228

work space

    flexibility in, 8

    reducing, 10

written communications, 212–213

Notes